FIREFLY

MARS

OBSERVER'S GUIDE

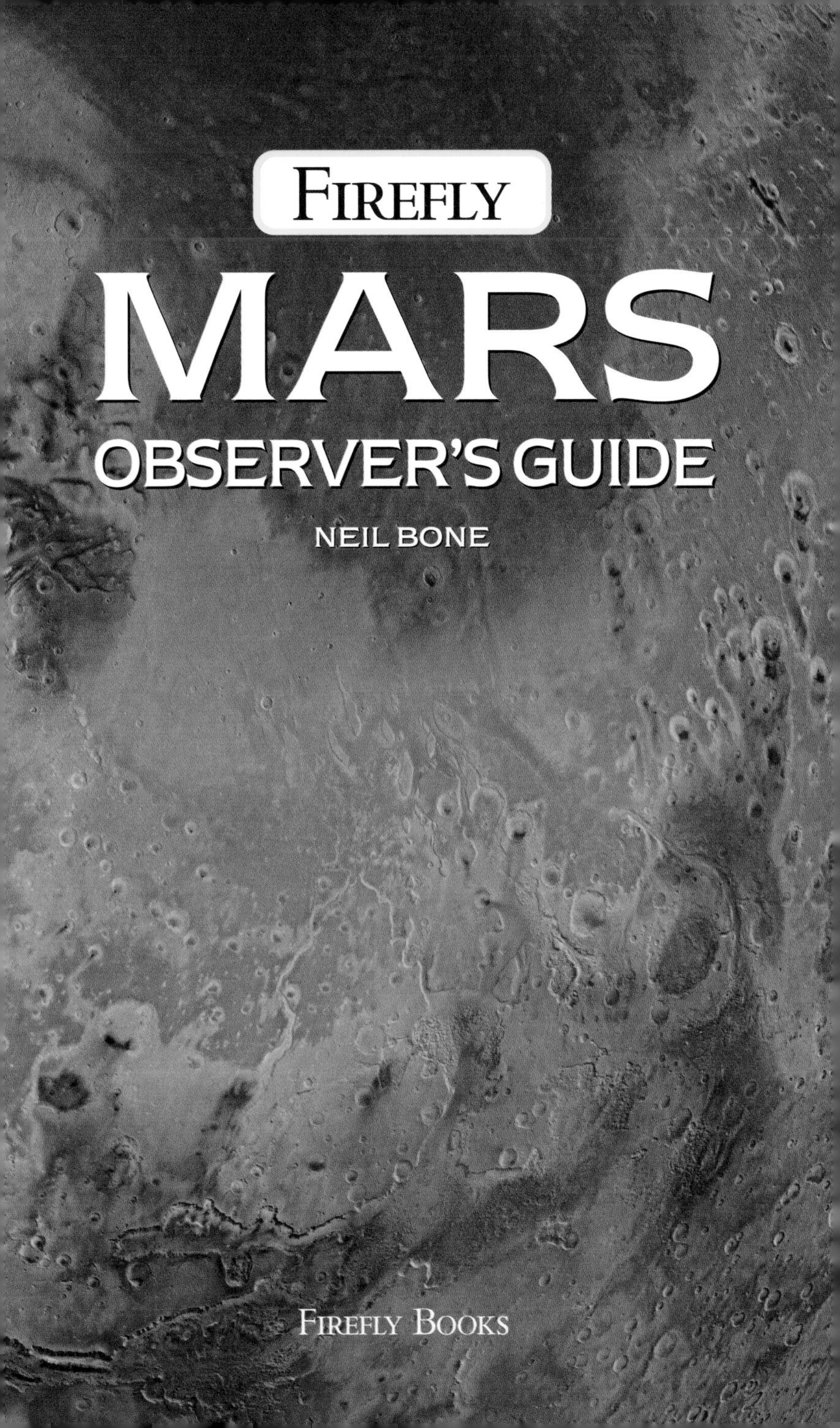

FIREFLY

MARS

OBSERVER'S GUIDE

NEIL BONE

FIREFLY BOOKS

For my daughter Miranda.

A FIREFLY BOOK

Published by Firefly Books Ltd. 2003

First printing

National Library of Canada Cataloguing in Publication Data

Bone, Neil
        Mars observer's guide : a practical handbook for amateur
astronomers / by Neil Bone.

Includes bibliographical references and index.
ISBN 1-55297-802-8

        1. Mars (Planet)—Observers' manuals. I. Title.

QB641.B65 2003   523.43   C2002-905553-9

Publisher in Cataloguing-in-Publication Data (U.S.)
(Library of Congress Standards)

Bone, Neil
        Mars observer's guide : a practical handbook for amateur
astronomers / Neil Bone.—1st ed.
[160] p. : col. ill. , photos. , maps ; cm.
Includes bibliographical references and index.
Summary: A guide to observe Mars and the various methods of recording
observations, from sketches to advanced imaging. Also includes the history
of the observation of Mars, the planet's structure and physical features.
ISBN 1-55297-802-8 (pbk.)
1. Mars (Planet)—Exploration. 2. Mars (Planet)—Observations. I. Title.
629.43/ 543 21   TL799.M3B66 2003

Published in Great Britain in 2003 by Philip's,
a division of Octopus Publishing Group Ltd,
2–4 Heron Quays, London E14 4JP

Published in Canada in 2003 by
Firefly Books Ltd.
3680 Victoria Park Avenue
Toronto, Ontario, M2H 3K1

Published in the United States in 2003 by
Firefly Books (U.S.) Inc.
P.O. Box 1338, Ellicott Station
Buffalo, New York 14205

Star maps (pp. 86, 95 and 102) by Wil Tirion.

Printed in China

CONTENTS

FOREWORD

The year 2003 saw an unusually close opposition of Mars, the closest since telescopic records began, and certainly the best for very many centuries to come. As Neil Bone explains in this topical book, at such times the Red Planet always becomes the center of attention for enthusiastic amateur astronomers.

No longer only the province of the science fiction writer, Mars is a real world whose daily exploration by orbiting craft such as Mars Global Surveyor can be followed in popular magazines, scientific journals and in real time over the Internet. Gone forever are the geometrically perfect "canals" of Giovanni Schiaparelli and Percival Lowell; gone forever are the intelligent, dying Martians of sci-fi myth. But new controversies continue to arise to keep the "Mars fever" alive. Even without the highly contentious (and ongoing) question of fossil bacteria, Mars will always remain an exciting world with dynamic meteorology, dust storms, constantly changing polar caps, huge volcanoes and a great rift valley. As a potential future base or second home for humankind its importance cannot be overstated.

Hopefully, after reading this book, you will at least want to observe Mars for yourself; perhaps just to follow its nightly movements against the starry backdrop, but *maybe* to take one further step. Amateur astronomers can play a part in gathering scientific data to support space missions. Long-term Earth-based studies provide a database of meteorological and other information, from which astronomers can build hypotheses and test theories. New observers are always needed, and modest telescopic equipment (whether used visually or with a CCD camera) can provide valuable data. *You* could be one of those observers!

Finally, whatever your level of interest or expertise, good luck with viewing Mars.

Richard McKim
Director, BAA Mars Section

—— *OBSERVING THE RED PLANET* ——

For as long as people have watched the skies, the naked eye planets have been objects of fascination. Brilliant Venus is well known for its alternate apparitions as an Evening or Morning "Star," while giant Jupiter is among the most prominent objects in the night-time sky when favorably placed. Mercury is more elusive, always staying close to the Sun. Saturn, taking a stately 29 years to complete its passage around the constellations of the zodiac, is comparatively dim to the naked eye by virtue of its distance, and is revealed in its true magnificence only when viewed through a telescope.

Of the five naked eye planets – literally "wanderers" from the Greek word describing their changing positions relative to the fixed star background – Mars is surely the one that has most fired the human imagination over the millennia. Its pronounced red coloration invites association with blood and war, and when relatively close to Earth, as at the opposition of 2003, Mars can outshine all else in the night sky except the Moon and Venus.

Among the other planets, only Venus comes closer to Earth than does Mars. For all their brightness, however, none of the planets reveals any detail to the naked eye, so knowledge of the planets' surface properties had to await the development of the telescope in the early 17th century.

Early telescopes were of poor optical quality, certainly in comparison with the instruments manufactured today. In the most primitive telescopes, Jupiter, with its family of four bright satellites, and Venus, with its changing Moon-like phases, proved the most rewarding planetary targets. Mars' surface details are rather subtle, and investigation of these had to wait until the development of better optics. In time, however, Mars came to be understood as a rocky world, similar in many respects to our own Earth, with dark surface markings and bright, white polar ice caps.

Mapping Mars' dark markings became a major activity for generations of planetary astronomers, who found that the markings vary in appearance over time. Mars has an atmosphere in which clouds can form, and where winds can whip up the surface dust into occasional vast, obscuring storms. Clearance and re-deposition of dust accounts for the ever-changing outlines of the dark features.

For much of the time during which it has been closely studied, Mars has been principally a target for visual observation. Until quite recently, most telescopic records have been obtained by the simple expedient of drawing what was visible in the eyepiece! Photography naturally came to the fore in the 20th century, but it was always lim-

▲ *The Wide Field and Planetary Camera (WFPC 2) on the Hubble Space Telescope has obtained some stunning full-disk images of Mars. This image, which was taken on March 10, 1997, just before Mars reached opposition, resolves features down to 22 km (14 miles) in size. The small north polar cap is visible at top, with the dark Syrtis Major prominent near the middle of the disk. At the south (bottom), the great Hellas basin is seen to contain haze. Cloud is visible on the right of the disk at Mars' evening terminator.*

ited by the unsteadiness of Earth's atmosphere, which tends to blur the image during the necessary long exposures. In the last 20 years or so, electronic charge-coupled device (CCD) detectors have largely replaced photography for many astronomical imaging purposes, bringing with them the advantage of shorter, sharper high-resolution exposures. Amateur astronomers are now able to produce CCD images of Mars that show the planet in more detail than the best professional photographs of, say, the 1960s.

◄ *Lowell crater is an impact structure in Mars' southern hemisphere, at longitude 081° and latitude 52°S. The crater has a diameter of 201 km (125 miles). It is seen here in an image from the Mars Orbiter Camera (MOC) on Mars Global Surveyor, taken on October 17, 2000. Frost is visible on the crater floor.*

Professional observatories continue to monitor Mars, and their work is further augmented by imaging from the Hubble Space Telescope (HST), which has, over the years, produced numerous superb full-disk pictures of the Red Planet. Orbiting above Earth's murky and turbulent atmosphere, HST is able to resolve extremely fine details.

Telescopic study of Mars by ground-based observers, both professional and amateur, remains a valuable activity. There are many amateur astronomers in organizations such as the British Astronomical Association (BAA) or the US-based Association of Lunar and Planetary Observers (ALPO) who dedicate most of their observing time to studying Mars when it is favorably placed. Working in cooperation, amateurs may even be the first to spot important events, such as the outbreak of a Martian dust storm, thus providing the alert so that professional cameras can be brought to bear. At times like the favorable presentation of the planet in 2003, this "hard core" of specialist observers is, not surprisingly, augmented by thousands of others with a more casual interest, who are keen to enjoy the best-possible views of Mars.

Our understanding of the Red Planet has accelerated since the dawn of the Space Age, with the first probes being sent to Mars in the early 1960s. Many surprising results have been obtained since then, including the discovery of geological features like the spectacular "Mariner Valley," giant volcanoes and extensive areas of impact cratering. These features remain beyond reach of even the best Earth-based telescopes, and it is obvious that a truly detailed knowledge of Mars requires study by spacecraft in orbit around the planet, and the use of landers to investigate conditions on the surface.

Part of the fascination of Mars, of course, lies with the possibility that conditions early in the planet's geological history may have been favorable for the appearance of life. The chance that Mars might at one time have been – or may even still be – the abode of simple microorganisms commands a lot of attention among some scientists and the general public. In 1976 NASA attempted to address the question of whether simple life exists in the Martian soil with experiments aboard the two Viking landers. Depending on one's viewpoint, the results were either negative or inconclusive.

▼ The surface of Mars as seen from the 1976 Viking 2 lander, which came down on Utopia Planitia, appears as a boulder-strewn desert. A cover, which was ejected from one of the lander's instruments, lies in the foreground. Visible at center is a trench that was dug by the scoop on the lander's robotic arm during the collection of soil samples for analysis.

This is a debate that refuses to go away! The famed "Martian Meteorite" ALH 84001, proposed by some scientists in 1996 to contain fossil bacteria, renewed interest in the possibility of life on Mars, and the question may not be resolved to the satisfaction of all until further direct sampling of the planet's soil is carried out during future spacecraft missions.

Still further in the future lies the possibility of manned exploration of Mars. After the Moon, the Red Planet was to be the next target for astronauts to collect samples from and conduct scientific experiments on the surface of another world. In the 30 years since the end of the Apollo program, speculation as to when Mars might be visited by a manned mission has veered from the optimistic to more realistic assessments. Such an undertaking – which will doubtless eventually come, given mankind's inquisitive nature – is fraught with hazard: interplanetary space is a more dangerous environment for the necessary long-duration trip than is low orbit around Earth. The colossal expense of a manned Mars mission is also no minor consideration. It may be several decades yet before we can do more than view Mars from a distance, or second hand through the instruments of automated orbiters and landers.

Scarcely a week goes by without at least one news story relating to Mars, be it a report on the latest planned spacecraft exploration, further speculation on the possibility of a warmer, wetter past – more amenable for life! – for the planet, or new analysis of the many images returned by Mars orbiters.

Mars myths

From its pronounced, angry red hue, it is hardly surprising that Mars has been associated since ancient times with conflict. Our modern name for the planet is that of the classical Roman God of War, equivalent to the Greek Ares, symbolic of strife and turmoil in human relationships. The Greek root is commonly used in describing features or characteristics of the planet – hence "areological" in the same sense as geological, or "areographic" and so forth. In Greek mythology, Ares was the son of Zeus and Hera. While one of the immortals – with a seat on Mount Olympus, upholstered in human skin! – it would appear that he was far from unbeatable, having been defeated in battle by Heracles, among others.

Mars' war-like association apparently goes back even further. The Chaldean skywatchers of more than 6000 years ago saw the Red Planet as an embodiment of Nargel, who was the God of Battle and God of the Dead. The ancient Persians had the planet as the Celestial Warrior.

Many of the ancient cultures who watched and recorded the movements of the planets did so for astrological, rather than astronomical, purposes. The Babylonians of 2000 BC, for example, left behind cuneiform tablet records of their observations of Mars and the other planets. The Mayans of South America also followed the planets' motions and must have been aware of Mars' movements relative to the "fixed" star background. However, most of the Mayan astronomers' attention appears to have been focused on the cyclical appearance of Venus, from the few written records to survive the conquistadors.

Following on from the Hippocratic tradition originating in classical Greece, Arab physicians of the 12th century based their treatments partly around a balance of "humors" or bodily fluids – yellow bile, black bile, blood and phlegm. Correctly balanced in the body, the humors would give good health, while imbalance would cause illness. To a degree, astrological thinking impinged on the belief system underlying medical thought at this time, such that slow-moving, somber Saturn was seen to influence the amount of black bile, leading, for example, to depression. Mars, acting through yellow bile, brought about anger.

Prominence of Mars in the sky, or indeed the appearance of comets or any number of other celestial phenomena, have always been taken by those who wish them to be as harbingers of strife or doom. The influence of astrology may be rather less in our hopefully more scientifically enlightened times, and most would look to Mars' brightness in, say, 2003 or 2005 as an indication of the planet's favorable positioning for telescopic viewing, rather than interpreting it as a portent of war!

Mars has always been the subject for some rather tangential thinking. The suggestion that some of the telescopic features might be linear channels (*canali* in his native Italian) by Giovanni Schiaparelli in 1877 was, for example, taken to the extreme by some other astronomers of the day. To some, *canali* rapidly translated to canals, put in place to irrigate a dry, dying planet by a desperate civilization. Over subsequent decades, the perceived network grew ever more complex in the eyes and drawings of some observers, before the existence of the canals was shown to be illusory in the 1900s. No trace of the canals of popular imagination was found when Mars began to be mapped in detail by orbiting spacecraft, of course!

Around the time of Schiaparelli's announcement of linear features on the planet's surface, Mars was discovered to have two moons. It is often pointed out that the English novelist Jonathan Swift alluded in his *Gulliver's Travels* (1726) to the inhabitants of his fictional Laputa being aware of Mars' possession of two moons, with orbital characteristics reasonably close to those of Phobos and Deimos, well prior to their discovery. The most satisfactory explanation is surely coincidence rather than conspiracy or unaccountable foreknowledge.

The era of mapping Mars from orbit has engendered its own latter-day mythology. The Mariner, Viking, Mars Odyssey and other spacecraft have produced hundreds of thousands of excellent images of Mars' surface, but the single most-often reproduced picture must surely be that taken in 1976 by the Viking 1 orbiter, showing what soon came to be known popularly as the "Face on Mars." Located in the Cydonia region, which is part of the flat expanse of the Acidalia Planitia (plain) in Mars' northern hemisphere, this rocky outcrop some 3 kilometers (2 miles) long showed, under the illumination conditions of the image, what appeared to be a mouth, nose and two eyes. Scientists were quick to point out that this was simply an illusion of light and shadow, but to the gullible, this has always been the ultimate conspiracy: clearly, in their eyes, the scientific

▼ *Shown here is the Viking orbiter image of the Cydonia region taken in 1976. Under the solar illumination conditions at the time, the mesa (table mountain) at top center gave the appearance of the "Face on Mars." This image became an icon for conspiracy theorists, believers in alien civilizations and film-makers alike.*

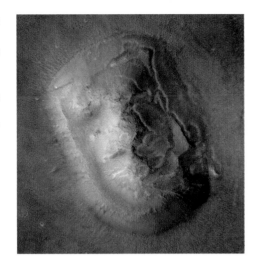

▶ Detailed imaging from Mars Global Surveyor in April 2001 confirmed the "Face" on Cydonia, in the Chryse Planitia region of Mars, to be nothing more than an illusion of light and shadow on a natural rock formation.

establishment was trying to conceal the existence on Mars of a giant monument erected by a Martian civilization.

Setting aside the obvious counter that were NASA worried about revealing this then surely the picture would never have entered the public domain, several images of the same feature – an eroded low-lying rock mesa (table mountain) on the Martian plains – have shown it to be nothing more than a trick of the light. Images from the orbiting Mars Global Surveyor spacecraft taken in April 2001 conclusively show that there is nothing in Cydonia beyond a natural rock formation. Nonetheless, the "Face on Mars" remains an icon for those who wish to cling to the idea of an advanced civilization having occupied the Red Planet in the recent past.

All too often in the case of Mars, it is easy for the naive to arrive at the remarkable explanation, when the mundane will suffice. Telescopic observers of the Red Planet have often learned the hard way to be cautious in interpreting what they see, but it is harder to make the conspiracy theorists adopt the same discipline!

Modern myths – Mars in science fiction

Mars has, hardly surprisingly, been the inspiration for countless science fiction stories. Perhaps the best known is H.G. Wells' 1898 classic *The War of the Worlds*. Taking his cue from the suggestion by the American astronomer Percival Lowell and others that the alleged Martian canals might represent the struggle of a beleaguered Martian civilization to irrigate their dying planet, Wells develops a scenario in which invaders from the Red Planet arrive to take over

▶ At opposition in June 2001, Mars – seen here at lower left – was relatively close in the sky to its "rival," the star Antares (lower right). Mars was by far the brightest object apart from the Moon in the night sky at this time, as it will be again in 2003.

the more habitable Earth. The tale has been adapted for motion pictures on countless occasions and, famously, triggered widespread panic when used by Orson Welles as the basis for a radio broadcast in the United States on October 30, 1938, with many listeners taking the drama to be a real Martian invasion.

Mars has served as the backdrop for numerous science fiction stories, including the 1920s pulp yarns of Edgar Rice Burroughs, with the planet under the guise of Barsoom. Others who have set stories on Mars include Arthur C. Clarke (*The Sands of Mars*, 1951), Robert Heinlein (*Red Planet*, 1949), Greg Bear (*Moving Mars*, 1993), and even Patrick Moore (a series starting with *Mission to Mars*, 1955). Perhaps most eloquent among the science fiction writings based around the Red Planet and its arid landscape are Ray Bradbury's *Martian Chronicles* (compiled in 1950), which dwell less on scientific realism and more on the human response to the environment and its doomed inhabitants.

More recent authors have focused on the possibilities for terraforming Mars for human habitation, as in the series *Red Mars*, *Green Mars* and *Blue Mars* (1992–6) by Kim Stanley Robinson. It is

not difficult to see a parallel between Mars and the desert world of Arrakis portrayed in the 1965 novel *Dune* by Frank Herbert.

Mars has also been popular as a setting for motion pictures. In television's *Star Trek*, the Federation has a dry dock in Utopia Planitia. Movies based around the manned exploration of Mars have included *Capricorn One* (1978) – surely one to satisfy the conspiracy theorists! – and *Mission to Mars* and *Red Planet* (both 2000, and unrelated to earlier novels of the same titles). Much of the background imagery in these fictional versions of Mars is drawn from the real panoramic views returned from the surface by the Viking and Mars Pathfinder landers.

With the Red Planet a beacon in the skies of August 2003, November 2005 and December 2007, Mars will be the subject of much popular interest in the years ahead. This book aims to present a basic guide for those wishing to observe Mars for themselves at these favorable apparitions: within the limitations of our telescopic equipment and a rather distant view, we can each be explorers of this fascinating world.

MARS AS A PLANET

The Solar System can broadly be divided into two "domains" – the inner, terrestrial planets (Mercury, Venus, Earth and Mars, in order of increasing distance from the Sun) and the outer, gas giants (Jupiter, Saturn, Uranus and Neptune). Between the two lies the asteroid belt, a collection of small, mostly rocky bodies which were prevented from coalescing into a single Moon-sized world by the influence of Jupiter's gravity early in the Solar System's history. On the Solar System's outer fringes lies the Edgeworth–Kuiper belt, comprising icy planetesimals, which are regarded by many astronomers as giant comet nuclei. These bodies are, like the asteroids, remnants from the formation of the Sun's planetary family. Pluto, outermost of the recognized planets, may simply be the largest of this population of icy bodies.

The distinct domains across the Solar System probably reflect the differing conditions within the protosolar nebula from which the Sun and planets condensed some 4.6 billion years ago. In the nebula's inner parts, where higher temperatures prevailed, only refractory materials like metals and silicates could condense, leading to rocky planetesimals which eventually coalesced to form the terrestrial planets. Farther out, where it was cooler, growing protoplanets were able to accumulate more volatile material from the surrounding nebula, leading to gas giant planets like Jupiter and Saturn.

Mars is the outermost of the terrestrial planets, and among these it is in many respects the most similar to our own Earth. Heavily cratered innermost Mercury, close to the Sun and with negligible atmosphere and an oversized metallic core, bears little resemblance to Earth beyond its rocky nature. Venus, while close to the same size as Earth, shows evidence for wholesale volcanic turnover and is shrouded in a thick, corrosive atmosphere in which a runaway greenhouse effect produces surface temperatures of 480°C (900°F).

Orbit

Mars orbits the Sun at a mean distance of 227,940,000 km (141,640,000 miles), or 1.52 AU. (One Astronomical Unit, AU, is equivalent to the Earth's mean orbital distance from the Sun of 149,597,970 km/92,960,000 miles.) Mars' orbit is markedly elliptical, so that at its closest – a point known as perihelion – it lies 206,700,000 km (128,400,000 miles/1.38 AU) from the Sun. Aphelion, its most distant orbital point, occurs at a distance of 249,100,000 km (154,800,000 miles/1.67 AU). As we shall see later, this substantial difference in orbital distance from the Sun has a marked influence on Mars' seasons.

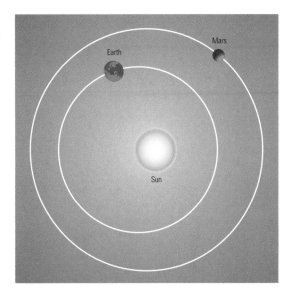

▶ The orbits of Earth and Mars are compared. Earth's is more or less circular but that of Mars is markedly more elliptical, with a considerable difference in distance from the Sun between perihelion (the point in the planet's orbit when it is closest to the Sun) and aphelion (the point when it is most distant). The orbits are drawn to scale, but the sizes of the Sun and planets are exaggerated.

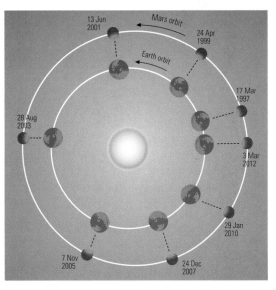

▶ Thanks to the marked ellipticity of Mars' orbit, some oppositions are more favorable than others. In August 2003, opposition occurs when Mars is close to perihelion, and the planet will appear bright and show a comparatively large telescopic disk. When opposition occurs near aphelion, as in 1997 or 2012, Mars is fainter and appears smaller.

Being farther from the Sun, Mars takes longer than Earth to complete an orbit – 687 days, or 1.88 Earth years. Earth and Mars line up so that the Red Planet is directly opposite the Sun in our sky once every 25–26 months. The precise interval varies considerably, depending on whether Mars is close to perihelion or aphelion and, therefore, moving more rapidly or slowly. Around the time of opposition, Mars is at its closest to Earth in any given apparition (visibility period). As a result of the elliptical shape of Mars' orbit, some oppositions bring the planet much closer to Earth – which has a more nearly circular orbit – than others. If opposition occurs at a time when Mars is close to its perihelion, the planet will appear particularly bright and will show a comparatively large disk when viewed telescopically. Such a perihelic opposition occurs in 2003. Mars will then be 56 million km (35 million miles/0.37 AU) from Earth. These oppositions recur at intervals of 15 to 17 years, with those of 1971, 1988 and 2018 also being perihelic. On such occasions, Mars outshines every other body except the Moon and Venus in the night sky. The oppositions of 2005 and 2007–8 also bring Mars reasonably close. Conversely, when Mars is close to aphelion at opposition – as in 1980, 1995 or 2010 – it appears relatively small and is not particularly prominent except by virtue of its pronounced red color. At these times, Mars becomes as bright as Sirius – most brilliant of the stars – but only about half as bright as Jupiter. Mars at aphelic opposition gets no closer than 100 million km (60 million miles/0.67 AU) from Earth.

Looping the loop

The superior planets, those in orbits beyond that of the Earth (in contrast with the inferior planets, Mercury and Venus, whose orbits are closer than ours to the Sun), show a broadly similar visibility pattern. At the point where they are at the same longitude on the sky as the Sun, they are said to be at conjunction, and they are lost in the Sun's glare. After conjunction, the apparition begins, with the planet emerging into the predawn sky west of the Sun, rising earlier each night. Eventually, at opposition – 180 degrees in the sky from the Sun – the planet rises at sunset, reaches its highest in the sky at midnight, and sets at dawn. After opposition, the planet rises before sunset and sets before dawn, with its visibility eventually becoming restricted to the evening hours. The apparition comes to a close when the planet is swamped by the evening twilight glare to the east of the Sun and conjunction is reached once more.

The orbits of the major planets (excepting Pluto) all lie within a relatively flat plane, 14 degrees wide. Projected on to the sky background, Earth's orbital plane – defined by the Sun's apparent annual path around the stars – marks out the ecliptic. The ecliptic is a great circle on the imaginary celestial sphere surrounding the Earth; it runs through the constellations of Aries, Taurus, Gemini, Cancer, Leo, Virgo, Libra, Scorpius, Sagittarius, Capricornus, Aquarius and Pisces. Since all the other planets have orbits close to the ecliptic

► Mars' retrograde "loop" motion relative to the background stars perplexed ancient astronomers but has a relatively straightforward explanation. For most of the time, the planet appears to move eastward from night to night. In the weeks around opposition, however, Earth speeds past on its smaller, shorter orbit and overtakes Mars. During this interval, Mars appears to move "backward" (retrograde). As the distance between the two planets increases again after opposition, Mars' eastward motion against the stars resumes.

plane, they are always found somewhere among these constellations. Mars' orbit is inclined to the ecliptic by just under two degrees.

Over the course of its orbit, each planet moves eastward along the ecliptic relative to the background stars. The more distant, outer planets move more slowly than the inner planets – Jupiter takes 12 years to complete an orbit, and Saturn takes 29 years. Mars, however, moves more quickly, and is the planet that most obviously shows an apparent westward – or retrograde – motion around the time of opposition.

This apparent "backward" motion of the planets relative to the stellar background perplexed some of the ancient astronomers, who came up with elaborate schemes of epicycles – circles imposed on the planets' orbits, then believed to be centered on Earth rather than the Sun – to explain it. The true situation is, in fact, fairly straightforward. All the planets orbit in the same direction around the Sun – counterclockwise as seen from above the north pole. Planets close to the Sun orbit more rapidly than those farther out. Consequently, Earth will regularly "overtake" Mars around the time of opposition, when the two worlds are closest to each other.

In the months well ahead of opposition, when located in the morning sky, Mars appears to track steadily eastward. As opposition nears, however, the rate of eastward motion slows until, about a month before opposition, Mars comes to a halt then starts to move westward relative to the stars. Earth at this time is passing Mars on the inside. As Earth recedes, leaving Mars behind, the apparent retrograde motion slows and stops, then, about six weeks after opposition, direct eastward motion resumes. Plotted on a star map, Mars' path in the months close to opposition appears as a loop in the sky, as can be seen on the maps in chapters 5, 6 and 7.

Martian geology

Mars has a diameter, at its equator, of 6794 km (4222 miles), which makes it the third largest of the terrestrial planets, and about half the size of Earth. Like Earth, Mars is slightly flattened, with a polar diameter of 6759 km (4200 miles).

In common with the other terrestrial planets, Mars has a layered (differentiated) internal structure, with a small metallic core overlain by a mantle, which in turn is covered by an outer crust. The core is believed to have a maximum diameter of about 2000 km (1200 miles), while the mantle, comprising the bulk of the planet, is 2300 km (1400 miles) deep. Thanks to its relatively small size, Mars is thought by planetary scientists to have cooled considerably since its formation, meaning that there is less molten material in the mantle than in those of larger bodies such as Earth, which retain much of their primordial "heat of formation." Another consequence of Mars' small size is the thickness of its crust – 100 km (60 miles) compared with a maximum of 50 km (30 miles) for Earth's crust.

Earth's metallic core shows dynamo effects, perhaps a result of convection currents in its semiliquid outer parts; the absence of these currents in Mars' core means that the planet lacks an appreciable extended magnetic field.

The surface geology – or, more correctly, areology – of Mars shows a division into two halves. The planet's northern hemisphere is dominated by relatively smooth, boulder-strewn plains, while the southern hemisphere shows more ancient, cratered terrain, which bears the scars of meteorite impacts dating back to the early days of the Solar System. Proximity to the asteroid belt placed the early Mars in the firing line during the late heavy bombardment phase of planetary formation around 3.8 billion years ago. During this phase, the majority of the remaining large planetesimals in the inner Solar System were swept up by the essentially fully formed planets. Since that time, the cratering rate on the terrestrial planets has slowed considerably. Weathering has eroded the evidence of this stage of formation on Earth, but the Moon and Mars' southern hemisphere bear witness to its intensity.

▼ *Schematic cross-sections through Earth and Mars are shown to scale for comparison. Mars is just over half as large as Earth, and has undergone greater cooling since its formation,* *meaning that its crust is thicker. The apparent lack of recent volcanic activity suggests that – unlike that of Earth – Mars' upper mantle now contains little molten material.*

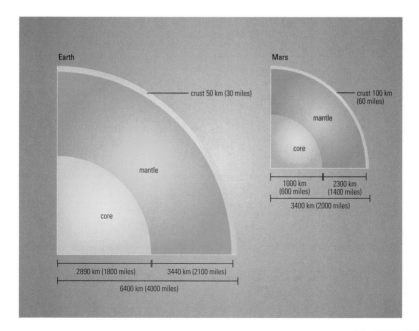

The northern hemisphere of Mars is, as a whole, lower in elevation relative to the mean and has fewer impact craters: weathering and resurfacing have erased the record of late heavy bombardment here. Cameras aboard the Viking and Pathfinder landers – each of

▼ This topographic map of Mars, based on Mars Global Surveyor data, shows the difference between the planet's two hemispheres. Relative to the arbitrary datum (taken to be the level at which atmospheric pressure is 6.2 millibars), the northern hemisphere is largely flat and low-lying (blue colors) while the southern hemisphere is more rugged and elevated (orange and red). The depth of the Hellas basin is very apparent.

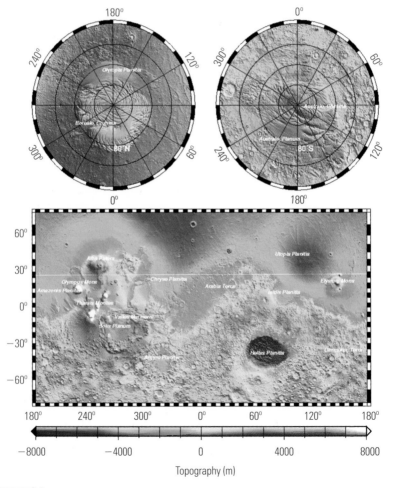

Topography (m)

which came down at sites in the northern hemisphere – recorded a landscape of sand and dust littered with rocks and boulders. The rocky material appears to be igneous (volcanic) in origin, and ranges from centimeter- to meter-sized blocks. The blocks are thought to be ejecta from impact events, thrown from their sites of origin, or perhaps carried to their current location by brief, violent water flows. Close examination shows the blocks to have pitted surfaces, the result of weathering.

Much of the Martian topsoil is made up of small dust particles, of the order of 10 micrometers (μm, a hundredth of a millimeter) in size, which can be easily blown around by surface winds. In places, sand-dune like accumulations are seen. The changing outlines of Mars' dark markings can be accounted for by varying levels of coverage of the underlying terrain by wind-blown dust.

The northern hemisphere's plains are extensive: Utopia Planitia (where the Viking 2 lander came down in 1976) has a diameter of 3200 km (2000 miles); Acidalia Planitia is similarly large; and the north polar cap is surrounded by the huge, flat Vastitas Borealis.

In situ chemical analysis of Martian rocks and dust shows these to have an abundance of iron-oxide containing minerals: Mars owes at least some of its ochre hue to rust!

Samples presumed to be of Martian rock have been examined on Earth, too. Around 20 meteorites studied by scientists are believed to be Martian in origin, having been thrown into interplanetary space by past impacts and subsequently swept up by Earth. Analysis of these SNC meteorites (the initials are for the type names Shergottites, Nakhlites and Chassignites, which are taken from their fall sites) reveals trapped gases with an isotopic composition identical to that of Mars' atmosphere as measured by the landers. The solidification ages of most SNC meteorites range from 150 million to 1.3 billion (thousand million) years ago, suggesting that Mars' volcanoes – from which these basaltic rocks are presumed to originate – were active comparatively recently in geological time.

Volcanoes

Although its mantle is relatively cool and its crust thick and rigid, Mars shows ample evidence for past volcanic activity. Some scientists even believe that this activity could still be ongoing at low levels. Dating of the SNC meteorites, supported by measurements of the impact cratering rate, suggests Mars' volcanic terrain to be as young as 100 million years in places.

Straddling the Martian equator around longitude 110°W is an elevated region called the Tharsis Ridge, which rises 27 km (17 miles)

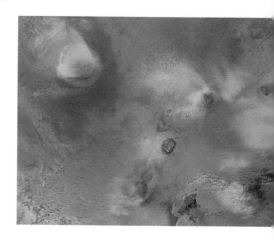

► *This Mars Global Surveyor image shows the Tharsis region volcanoes. Olympus Mons is at top left, while Arsia Mons, Pavonis Mons and Ascraeus Mons, respectively, run from bottom center to mid-right. Orographic cloud can be seen on the flanks of the giant volcanoes.*

above the surrounding terrain. Atop Tharsis sit three large shield volcanoes – Arsia Mons, Pavonis Mons and Ascraeus Mons – which can sometimes be seen as spots from Earth in a good telescope under exceptional conditions. To the northwest of Tharsis is a fourth volcano, Olympus Mons. Still further north and west is a fifth giant volcano, Elysium Mons, which tops a smaller, but still significant, bulge in the Martian crust.

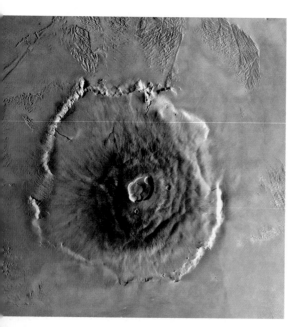

◄ *The largest of Mars' volcanoes – indeed, the largest in the Solar System – is Olympus Mons, towering to an altitude of 24 km (15 miles). This view is a composite of images obtained from the Viking 1 orbiter in June 1978, compiled as part of the Mars Digital Image Model.*

Rising to a summit height of 24 km (15 miles), Olympus Mons is the largest shield volcano in the Solar System. The closest terrestrial equivalent is Mauna Loa in Hawaii, which reaches 9 km (6 miles) from the ocean floor, making it a good deal smaller. Olympus Mons has an overall diameter of 700 km (430 miles), with gently sloping flanks leading up to a summit caldera (crater) 80 km (50 miles) wide. Again for comparison, Mauna Loa has an overall diameter of 120 km (75 miles).

Terrestrial volcanoes form above plumes of molten rock (magma) rising through the mantle. Plate tectonic motions carry regions of crust over these "hot spots" where they may remain for millions of years before moving on. The Hawaiian island chain developed as a result of this process: Big Island, as the current occupant of the position above the hot spot, is still growing, whereas others in the chain have become eroded over time. On Mars, however, the crust is immobile, meaning that eruptions above any magma plumes continued to build up lava flows in the same place over extremely long periods, which explains why Olympus Mons and the volcanoes of the Tharsis Ridge grew to vast size. Arsia Mons has a diameter of 800 km (500 miles) and stands 9 km (6 miles) high, while Ascraeus Mons towers to a height of 18 km (11 miles).

Faults

Mars' thick crust does not show the plate tectonic motions that shift Earth's continents over time. Instead, it appears to be a single unsegmented unit. There are, however, dramatic examples of faulting in the crust, particularly the huge Valles Marineris complex, which runs roughly parallel to the equator at 14°S and stretches east–west for some 4500 km (2800 miles). This vast canyon system, up to 600 km

▼ *Shown here is a composite image of the 4500-km-long (2800 miles) Valles Marineris canyon complex, to the east* *of Tharsis. Noctis Labyrinthus, an area of fractured terrain, is at the left (western) edge in this view.*

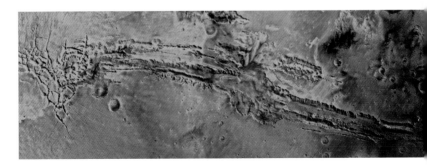

(370 miles) wide and as much as 7 km (4 miles) deep, dwarfs Earth's Grand Canyon (which is a mere 340 km/210 miles long, 11 km/7 miles wide and 1.5 km/1 mile deep). The "Mariner Valley" takes its name from the Mariner 9 spacecraft whose cameras first recorded it in 1972.

Located just east of Tharsis, the Valles Marineris may be the result of stress and pressure on the Martian crust caused by the enormous mass of erupted material in the adjacent volcanic region. The complex begins in the Tharsis ridge itself with an area of fractured terrain known as Noctis Labyrinthus, where numerous fairly narrow canyons intersect each other. Two broad chasms at the west – Ius Chasma and Tithonius Chasma – run parallel to each other, with the canyon system broadening and opening out in its central region. The eastern end of Valles Marineris is dominated by what planetary scientists call chaotic terrain, where the ground has collapsed. Features in Ius Chasma and in the chaotic terrain are indicative of geological processes involving water: the Mariner Valley is large not only because of faulting in the Martian crust but also because of water erosion and landslides.

Water outflow features

At Mars' mean surface temperature of −63°C (−81°F), the planet's water is expected to be mostly frozen out in permafrost in the soil, or in polar ice cap deposits. Spacecraft observations, however, have revealed numerous features that can only be dried riverbeds and channels. These features have been taken by many scientists to indi-

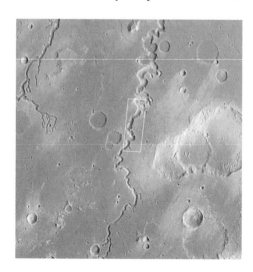

▶ *This Viking I orbiter image shows sinuous water flow features in the Xanthe Terra region of Mars.*

cate that there have been times in the past when Mars was warm enough, and the atmosphere sufficiently dense, to support running water – at least temporarily – on the surface. Such features are found most notably in the region northeast of the Valles Marineris complex, breaking out of the chaotic terrain into the flat expanse of Chryse Planitia (in which the Viking 1 lander set down in 1976). Outflow channels are also found among the southern hemisphere's craters.

In parts of the Valles Marineris, the canyon walls show stratification consistent with the deposition of sediments in lakes of standing water. One possible explanation for some of the water outflow features around the Valles Marineris lies with the sudden collapse of natural ice or rubble dams retaining such lakes, causing cataclysmic, short-lived flooding events.

At the present epoch, Mars is certainly arid, and there is nothing to suggest that water is found today on its surface. There is, however, ample evidence for subsurface deposits. Mars has a considerable reservoir of water frozen into its permafrost, particularly at higher latitudes in the southern hemisphere, as detected by the neutron spectrometer instrument aboard the Mars Odyssey spacecraft inserted into orbit around the planet in October 2001. The quantity of water ice bound up in this permafrost, less than a meter below the surface, may be sufficient to have allowed the formation of extensive lakes or seas in the distant past, which would account for features of apparent sedimentary origin revealed in several locations. One estimate suggests that if all the bound-up water were to be released Mars would be covered by a global ocean to a depth of as much as 500 meters (1600 feet).

The release of subsurface ice deposits, liquefied by pressure, offers a good explanation for many of the observed outflow features. Crustal movements caused by faulting, or in response to large meteorite impacts, might have forced groundwater to break through and briefly flow on and just below the planet's surface. Such a mechanism accounts for the shortness of some flows, and their brief lifetimes: many of Mars' dried riverbeds lack the extensive branched tributaries that would have formed if the flow had been of long duration.

Rather than postulating a warm, wet past for Mars, many geologists are happier to attribute the numerous outflow features to repeated, fairly frequent episodes of groundwater breakout over an extended period of the Red Planet's history. Association of the water flow features with chaotic terrain lends weight to this hypothesis. In regions such as Ius Chasma, groundwater outbreaks could have been sufficiently frequent as a result of landslide events to form lakes, thus explaining the observed layered sediments.

Craters

One of the early surprises of the Space Age was the discovery that large parts of Mars' surface were heavily cratered: prior to the Mariner 4 fly-by in 1965, it had been assumed that the planet had a relatively smooth surface dominated by rock outcrops and wind-blown sand. None of Mars' craters is sufficiently large to be visible as such in Earth-based telescopes – the largest are no more than 400 km (250 miles) in diameter.

Most of Mars' craters are found in the ancient, higher-elevation terrain in the southern hemisphere. Like the vast majority of those on the Moon, Martian craters are the product of high-velocity impacts of small bodies. On impact, the meteorites that produced the craters were completely vaporized. Craters are typically surrounded by a blanket of material (ejecta) thrown out during the impact. In general, the crater has a diameter about ten times that of the impacting body. Close to the crater's elevated rim, the ejecta blanket is dominated by larger blocks of material, with smaller material being thrown farther from the impact site.

The ejecta blankets surrounding Martian craters often show significant differences from those around lunar craters. On the airless, dry Moon, small fragments of material could be flung farther from the impact site, unchecked by atmospheric drag. Younger lunar craters in particular are often surrounded by an extensive bright radial ray system – the large crater Copernicus is a good example. By contrast, impacts in Mars' ice-bearing soil often appear to have

▼ The cratered terrain in Mars' southern hemisphere, shown in this Viking orbiter composite image, is evidence for heavy meteorite bombardment in the Red Planet's past. Water outflow channels are also visible. This view covers the region from 5° to 9°S at longitude 312° to 320°.

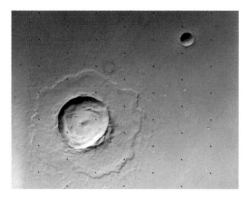

▶ Evidence for water, as permafrost in the Martian topsoil, is given by impact craters like Balz, on Chryse Planitia. The blanket of ejecta thrown out during the crater's formation has flowed as a semifluid mass to produce the observed lobate structure. The frame is 50 km (30 miles) across.

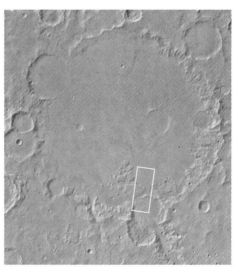

▶ Gusev crater, in the southern hemisphere, is a possible landing site for one of the Mars Exploration Rovers in 2004. The crater has a diameter of 150 km (90 miles), and shows features associated with the past presence of water (possibly even a lake), making it a potentially interesting site at which to search for traces of life.

produced ejecta with lobed flow structures at their outer edge, suggestive of temporary melting followed by re-freezing. Rather than being flung out as a mass of small particles, the finer ejecta from Martian craters appears to have flowed over the surrounding terrain as a muddy sludge.

Martian craters also show a greater degree of erosion than do those on the Moon. Frequent dust storms, volcanic activity (which was ongoing much later in Martian than lunar geological history) and water outflow processes have all taken their toll over time, degrading many of Mars' impact structures. Water outflow features are quite often found in the regions between craters in the planet's southern hemisphere.

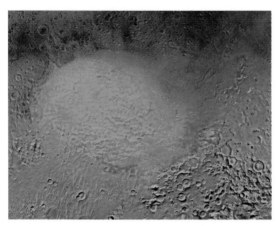

◄ This composite view of the 1800-km-diameter (1100-mile-diameter) Hellas basin was assembled from Viking orbiter images. Hellas is one of the Solar System's largest impact structures. It is is 5 km (3 miles) deep, well below the surrounding terrain in Mars' southern hemisphere.

Impact basins

While the common meteorite-derived impact craters on Mars escape telescopic detection from Earth, the scars of a couple of rather more sizable impacts are visible, though their true nature was not revealed until the Mariner 9 spacecraft began its detailed investigation of the planet's surface from orbit in 1971. Among Mars' most striking features is a huge circular impact basin, named Hellas Planitia, at 43°S latitude. Five kilometers (3 miles) deep, Hellas has a diameter of 1800 km (1100 miles) and is quite readily visible to Earth-based telescopes. The Argyre Planitia basin (at 50°S latitude), 900 km (560 miles) in diameter, is also visible from Earth. In addition, the surrounding topography strongly suggests that Chryse Planitia in the northern hemisphere occupies the extremely eroded site of a giant impact early in Martian history.

Both Hellas and Argyre are surrounded by concentric rings of fractures in the Martian crust. These ancient features result from the impact of large planetesimals during the later stages of Mars' formation. Similar, ancient, large impact basins are found on the Moon (the 2500-km/1600-mile-diameter South Pole–Aitken basin is the largest) and Mercury (the 1300-km/800-mile-wide Caloris Planitia).

Atmosphere

Partly as result of its small size and mass – and, therefore, low surface gravity (37.9% that of Earth) – Mars has been unable to retain a substantial atmosphere. Atmospheric pressure on the surface of Mars is of the order of 6 millibars, about 1/150th that at sea level on Earth, or equivalent to that found in the terrestrial stratosphere at an altitude of about 35 km (22 miles).

Mars' atmosphere comprises 95% carbon dioxide (CO_2), 2.7% nitrogen (N_2) and 1.6% argon (Ar), with the remainder being made up of trace quantities of oxygen, carbon monoxide and water vapor – certainly not a breathable mixture by human standards!

The current Martian atmospheric density is so low that any free water on the surface would rapidly evaporate: pools of standing water or long-lasting river flows are simply not possible. The existence of ancient water outflow features indicates that the atmosphere has been denser in the past. The 0.03% proportion of water vapor is, however, sufficient to allow formation of thin clouds and fogs. Spacecraft orbiting the planet have imaged fog forming in the Valles Marineris and in depressions such as the Hellas and Argyre basins. From the surface, the 1997 Mars Pathfinder lander imaged tenuous water ice clouds forming in early morning and evening at altitudes of 10 to 15 km (6 to 9 miles). They are similar in appearance to the noctilucent clouds seen in summertime in Earth's high atmosphere. Earth-based observers can sometimes see the orographic clouds produced when moisture-bearing Martian air is forced to rise and cool over elevated features such as the Tharsis volcanoes. Hazes can occasionally be seen close to the dusk or dawn terminator (day–night line).

Although largely composed of the greenhouse gas carbon dioxide (the abundance of CO_2 is twenty times that in the terrestrial atmosphere), Mars' atmosphere is far too tenuous to offer any substantial warming to the planet. The mean surface temperature is a chilly −63°C (−81°F), which is only four degrees greater than would be

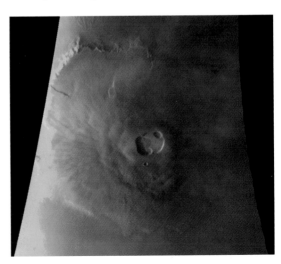

► Clouds can be seen toward the right in this image, forming over Olympus Mons, as seen from Mars Global Surveyor on October 20, 1997.

expected in the absence of an atmosphere. Earth, by contrast, is warmed by some 35–40°C (95–104°F) by its atmospheric blanket. On Mars, temperature minima can be as low as −137°C (−215°F), while summer temperatures may reach a relatively balmy +26°C (+79°F).

Planetary scientists have adopted the 6.2 millibar pressure level as the arbitrary datum – Martian "sea level" – for defining altitudes on Mars. Local pressure conditions are strongly influenced by topography. The floor of the Hellas basin, for example, is well below datum, while the peak of Olympus Mons reaches well above: atmospheric pressure at the top of Olympus Mons is about a factor of eight less than that in the depths of Hellas.

Seasons

Mars experiences seasons because its axis of rotation, like Earth's, is inclined to its orbital plane. Indeed, the current tilt is quite similar for the two planets, being 23°26′ for Earth and 23°59′ for Mars. Unlike Earth, Mars does not have a prominent northern Pole Star: seen from the Martian surface, the stars appear to wheel around a relatively empty patch of sky midway between Deneb in Cygnus and Alderamin in Cepheus.

When they are presented toward the Sun during summer, the respective hemispheres receive greater insolation and, therefore, reach

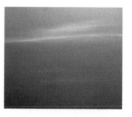

▼▶ (Right) Tenuous clouds in Mars' atmosphere were imaged from the surface by a camera on the 1997 Mars Pathfinder lander just before dawn. These clouds bear some similarity to the noctilucent clouds (below) that occasionally form at great heights in the Earth's thin upper atmosphere.

higher temperatures. Mars' seasons last about twice the length as those on Earth because its orbit takes nearly twice as long to complete.

In contrast with those on Earth, there is a marked asymmetry between the hemispheres with respect to the seasons. This is a result of Mars' greater orbital eccentricity. Summer in the southern hemisphere occurs when Mars is close to perihelion (its nearest point to the Sun), while that in the northern hemisphere comes at aphelion. At perihelion, Mars is some 43 million kilometers (27 million miles) closer to the Sun, so that the southern hemisphere experiences a greater level of summer heating than does the northern. By the same token, winters – at aphelion – are more severe in the southern hemisphere, so this half of Mars undergoes a greater range of temperature extremes over the year. Northern summers last 21 days longer than those in the south; equally, the southern winter lasts 21 days longer than that in the north.

These seasonal differences are reflected to some degree in the polar ice caps. The longer, colder winters in the southern hemisphere result in accumulation of a more extensive cap, which is subject to more rapid heating and melting during the spring. Both caps are permanent, probably several kilometers deep, and are prominent telescopic features. They comprise a mixture of water and carbon dioxide ice, together with dust deposits. The northern cap is mainly water ice, overlying smooth terrain and surrounded by the extensive sand dunes of the Vastitas Borealis. The southern cap sits atop cratered terrain like that elsewhere in the same hemisphere. In the autumn and winter, the southern cap is enshrouded by clouds of carbon dioxide (CO_2).

During their respective spring and summer, the polar caps shrink, retreating to higher latitudes as CO_2 sublimes (passing from the solid ice phase straight to vapor) in response to solar heating. This shrinkage is irregular, sometimes leaving behind detached "islands" of ice on the equatorward edge of the cap. At maximum, the caps may extend 30 degrees in latitude from the poles, retreating back to within 5 degrees at midsummer.

Release of CO_2 from the polar caps during spring and summer results in seasonal pressure variations. In summer, the local atmospheric pressure can be 25% higher than in winter. Water vapor released at the same time enhances the possibility of cloud formation.

At some perihelia, summertime heating can lead to the development of extensive dust storms, commonly originating in the Hellas basin. On occasion, as in 1956 and 1971, the dust storm may become planet-encircling, shrouding Mars' features from view for weeks or months at a time, as winds blowing at up to 400 km/h (250 mph) whip up fine particles from the surface. During major dust storms, the suspended material in Mars' atmosphere can act to trap more solar radiation, leading to a slight degree of global warming, which in turn contributes to the storm's spread and longevity. More localized dust storms break out quite frequently, typically in low-lying areas like the Valles Marineris and Chryse Planitia.

Images from the landers show that there is usually some dust suspended in the atmosphere, raised by gentler winds. Direct measurements at ground level by the Viking landers showed wind speeds of between 4 and 40 km/h (2 and 25 mph), with noon the breeziest part of the day. Martian equivalents of "dust devils" have been recorded.

Rotation

A point of strong similarity between Mars and Earth is the length of the day. Mars takes 24 hours 37 minutes 22.6 seconds to rotate on its axis – not that much longer than Earth's 23 hours 56 minutes 04.1 seconds. The Martian day is sometimes known as the "sol."

Satellites

Mars has two moons – Phobos and Deimos, respectively "Fear" and "Terror," the sons of the God of War. Both satellites are extremely faint, and they are lost to most telescopes in the glare from the planet itself. They were discovered at the perihelic opposition of 1877 by the American astronomer Asaph Hall.

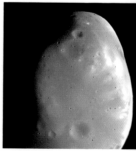

◀▼ *Mars' satellites Phobos (left) and Deimos (right) as imaged by the Viking orbiters. Phobos is the larger, and is marked by the 10-km-diameter (6-mile) crater Stickney, and strange grooved features. Deimos is also heavily cratered.*

Both satellites are tiny, and irregular in shape. Phobos is the larger, with dimensions of 27 × 22 × 18 km (17 × 14 × 11 miles). Deimos measures just 15 × 12 × 10 km (9 × 7 × 6 miles). The pair are held in synchronous rotation relative to Mars, which means that they always present the same face toward the planet, just as our Moon always keeps the same face toward Earth. They are tidally locked so that their longer axes are aligned to the planet.

Phobos and Deimos have fairly circular orbits close to Mars' equatorial plane. Phobos is much the closer, at a distance from the planet of 9378 km (5827 miles), while Deimos is 23,460 km (14,580 miles) away. A single orbit for Phobos takes 7 hours 39.6 minutes, meaning that it swings around the planet more than three times per Martian day. An observer on the surface would see Phobos rise in the west, then move quite rapidly across the sky to set in the east about six hours later. Deimos, by contrast, takes 30 hours 17.5 minutes (about one and a quarter Martian days) to complete an orbit, and would be seen from the surface to move slowly westward – opposite in direction to Phobos.

Phobos and Deimos are believed to be captured asteroids – a consequence of Mars' proximity to the main belt. Both are dark bodies, with albedoes of around 0.05: in other words, they reflect back only about 5% of the incident sunlight. Their slightly reddish coloration is similar to that of the so-called C-class asteroids found in the outer main belt. C-type asteroids are thought to be the parent bodies of the carbonaceous chondrite meteorites, rich in organic molecules, which occasionally fall on Earth. The satellites' overall densities of about 2 g/cm^3 are also consistent with the idea that they are captured asteroids; similar densities have been found for asteroids like Ida.

The Mars-orbiting Mariner and Viking spacecraft of the 1970s revealed much information about the physical natures of Phobos and Deimos. Images show them to be heavily cratered bodies, covered in a surface layer of pulverized rock fragments and dust (a regolith, similar to that covering the surface of the Moon and of the asteroid Eros). The cratering and regolith are products of repeated impacts by smaller, meteorite-sized bodies. House-sized boulders thrown out by the impacts lie on the moons' surfaces. Phobos is dominated by a 10-km-wide (6-mile-wide) crater called Stickney. Grooved stress fractures emanate from the crater, and it is possible that Phobos was almost shattered by the force of the impact that created Stickney.

Dissipation of tidal energy is causing Phobos' orbit to decay gradually. In 40 to 50 million years, Phobos will crash into the Martian surface. Some scientists contend that it will break up before it can do so, producing instead a ring around Mars. This would indeed seem a likelier outcome if Phobos has a loosely aggregate "rubble pile" structure similar to that of asteroids like Mathilde, rather than being a solid, monolithic body. Phobos' limited orbital future is again very much in keeping with the theory that it was captured by Mars, rather than forming in its neighborhood.

Mars is a dynamic world, with surface features that show subtle changes over time in response to wind-driven movement and deposition of dust. As we shall see later, some of these effects are visible to Earth-based observers. Temporary obscurations of features may occur because of dust or the condensation of ice clouds as the temperature changes. The ice caps shrink and grow with the seasons. In the course of an evening's observing, Mars' rotation will carry different features into view at the eyepiece. Over the course of a favorable apparition, such as in 2003, 2005 and 2007–8, Mars can be a most rewarding target for the patient observer, and no two views of the Red Planet are ever quite the same.

MARS DATA	
Mean orbital distance from Sun	227,940,000 km (141,640,000 miles)
Perihelion distance from Sun	206,700,000 km (128,400,000 miles)
Aphelion distance from Sun	249,100,000 km (154,800,000 miles)
Orbital eccentricity	0.093
Orbital inclination	1°50'59"
Orbital period	686.98 days
Equatorial diameter	6794 km (4222 miles)
Polar diameter	6759 km (4200 miles)
Axial tilt	23°59'
Rotation period	24h 37m 22.6s
Satellites	Phobos (maximum diameter 27 km/17 miles)
	Deimos (maximum diameter 15 km/9 miles)

EQUIPMENT BASICS

Compared with many astronomical targets, Mars has in its favor ease of visibility: its distinctive red color and brightness when close to opposition make Mars readily identifiable with the naked eye. Like the other planets, Mars requires telescopic examination before it will reveal any detail. At its greatest – as in August 2003 – Mars' disk subtends an apparent diameter on the sky of 25 arcseconds. For comparison, the Moon's apparent diameter of 30 arcminutes (half a degree) is over 70 times greater. All the naked eye will show, therefore, is Mars' position against the background stars. A careful observer might emulate the work of classical astronomers by plotting out the Red Planet's changing location over several months to detect its retrograde loop and note its changing brightness, but it will be possible to do little else.

Binoculars do little to improve the view! While the common 7×50 or 10×50 models can be recommended for many astronomical activities – deep sky and cometary viewing, or variable star work, for example – they do not provide sufficient magnification to present Mars as a discernible disk. Binoculars will reveal fainter stars in the background, but little else.

To resolve Mars' disk and the details on it, the observer will need a telescope. In terms of what can be resolved, the instrument's aperture – the clear diameter of its light-collecting lens or mirror – is all-important. Aperture governs both light-gathering ability and resolving power. A large aperture will show fainter stars and reveal finer detail. For instance, a 60 mm refractor should, in theory, resolve details on Mars which are separated by an angular distance of 1.9 arcseconds, while a 100 mm aperture instrument will resolve down to 1.2 arcseconds.

It is desirable when viewing a small planetary disk to resolve details as fine as possible, and in many respects a 60 mm refractor – a common beginners' telescope – will fare little better than binoculars as far as Mars is concerned. Such instruments will show Saturn's rings adequately when these are presented wide open (as in 2003), or the major dark belts of Jupiter, but they will struggle to bring out Mars' subtler details.

The ideal planetary telescope

The choice of telescope for visual planetary observing has long been a subject of debate among amateur astronomers. Three main telescope types are in common use – refractors, Newtonian reflectors and catadioptrics. Each has its adherents, and each also has its lim-

▼ The main telescope types are shown below. In a **refractor** (top), light is collected and brought to a focus by a front-end objective lens. An eyepiece behind the objective provides a magnified view of the focused image. The **Newtonian reflector** (middle) collects light with a large primary mirror. An angled flat secondary mirror diverts the reflected light out through the side of the telescope to an eyepiece mount perpendicular to the main mirror. In the popular **Schmidt–Cassegrain catadioptric reflector** (bottom), light is collected with a primary mirror, which reflects the light back on to a secondary mirror on the front-end correcting plate. The observer views the image through a central hole in the primary mirror.

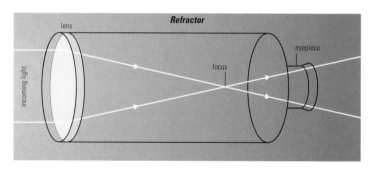

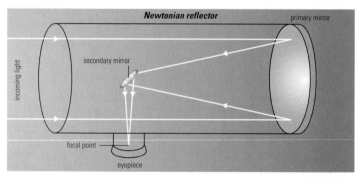

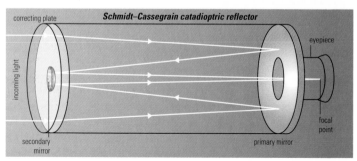

itations. At the outset, it must be said that it is sensible to make the best possible use of whatever instrument you have available: do not be deterred from at least trying to observe on the basis of others' negative opinions!

Regardless of optical layout, the most important feature of a telescope for astronomical use will generally be its aperture – that is to say, the diameter of the main glass element used to collect and focus light, be it a lens or mirror. Light-gathering power increases with aperture: for example, a 100 mm aperture instrument will show stars twice as faint as those at the limit of visibility in a 60 mm telescope. This is a major consideration for deep sky observers, but less so for planetary work. Where larger apertures do matter, however, is in their resolving power: large aperture instruments, within the constraints of atmospheric turbulence, allow finer details to be made out on a planet's disk (or, by the same token, more closely separated double stars, for example, to be resolved).

Refractors

Refractors are the "classic" telescope design, with a large front-end objective lens (sometimes called the object glass or OG) bringing light to a focus at the eyepiece end. The distance at which the eyepiece has to be positioned to meet the objective's focal point is described as the focal length. Dividing the focal length by the objective diameter gives the telescope's focal ratio, $f/$.

It is generally reckoned that to be useful for planetary observing a refractor needs to have an objective diameter of at least 80 mm, and preferably 112 mm. Refractors with apertures smaller than 80 mm lack the resolving power to bring out fine details, while the higher contrast afforded by larger instruments also improves the chances of being able to make out subtle variations in shading.

Refractors of up to 112 mm diameter are reasonably portable, and can be fairly easily and quickly set up on a garden tripod for an evening's viewing, followed by equally convenient dismounting and storage. Working at a fairly typical $f/8$, a 112 mm refractor will have a manageable tube length of 976 mm.

Once apertures of 150 mm and upward come into consideration, however, we enter the realm of substantial refractors with long tubes, and the requirement for a permanent mounting housed in its own observatory. Several professional observatories have such instruments, as do many run by local astronomical societies. Large refractors of this calibre are seldom manufactured nowadays, and many of the big refractors which still see service today have considerable histories attached to them. In the United Kingdom, the

6 inch (150 mm) refractors manufactured by Thomas Cooke and Sons of York in the 19th and early 20th centuries are cherished and much-admired instruments at many observatories, and still offer superb planetary views. In the United States, refractors from the same era by the Cambridge, Massachusetts, firm of Alvan Clark and Sons are similarly revered.

In recent years, there has been a vogue for producing small, very portable, short focal length refractors: 80 mm $f/5$ instruments (400 mm focal length), for example, are popular – I use one myself for recreational deep sky observing! These are sometimes described as "fast" (by analogy with camera lenses, where a small $f/$ number equates with shorter exposure) and offer good, contrasty views. Generally, however, they are not so well suited for planetary viewing as longer focal length systems, being less amenable to accepting high magnification eyepieces.

▼ An 80 mm long focal length refractor is suitable for planetary work. This telescope is on an altazimuth mount. The small finder telescope attached to the main tube offers low power views, enabling initial centering of viewing targets in the telescope's field.

▲ Viewing objects high in the sky with a refractor can involve an uncomfortable crouched observing position. One solution, which is also employed in SCTs, is to put the eyepiece in a diagonal which diverts the light by 90 degrees. A disadvantage of star diagonals is that they reverse the image left to right.

The major drawback of refractors is that unless extremely expensive optics are employed, the images they provide can be prone to false color. Refractors made with poor-quality glass will often show the bright limb (edge) of the Moon, or a planet's disk, to be fringed in blue. This results from *chromatic aberration*, a consequence of the unequal refraction (bending) of light at different wavelengths as it passes through a lens; red light is bent less than blue, with the result that these wavelengths come to focus at slightly different (but all too discernibly so!) positions. The effect can be alleviated to a large extent by the use of compound objectives, combining optically different glasses which compensate for each other. Refractors that do this are described as achromats. Still greater color correction is offered – at considerable cost – by apochromatic refractors.

A big problem for some observers is that refractors can demand uncomfortable viewing positions: when objects are high in the sky a refractor's long tube demands that the observer adopt a crouched position to access the eyepiece. One solution is the use of a right-angle "star diagonal" at the eyepiece end, which sends the image perpendicular to the aiming direction via a mirror. This, however, has the disadvantage of producing a laterally reversed (mirrored) image, making identification of features awkward relative to published maps. Perhaps the best solution is the use of a telescope mount that allows the observer to get under the focuser relatively comfortably: large refractors, particularly, are commonly set up on tall "pier" mounts, which give access to the viewing position without requiring the observer to crouch down.

Refractors for terrestrial use have an additional lens in the light-path to provide an upright image. This is unnecessary for astronomical use, but many observers do still use telescopes that give a "right-way-up" view. For astronomy, many would contend that the extra lens is merely an additional potential source of chromatic aberration and light loss, and is strictly something to be avoided. Whatever, there is no need for the telescope to provide an erect image for night-time viewing, and the additional lens will, of course, add to the cost of purchasing the instrument.

Many specialist planetary observers would say that their instrument of choice is a 112 mm apochromat, but it is still possible to make good observations with smaller, achromatic instruments.

Reflectors

The effect of chromatic aberration in refractors leads many observers to suggest that the Newtonian reflector, collecting and focusing light with mirrors instead of lenses, is preferable. Since light is reflected by the same extent regardless of wavelength, reflectors are free from chromatic aberration. As with refractors, reflectors come in a range of focal lengths; an instrument of about $f/6$ or greater is desirable for planetary work.

Unlike the "straight through" view in a refractor, a Newtonian reflector employs a focusing mount on the side of the tube, near the top end. Light from the primary mirror is reflected at right angles into the focuser from a smaller secondary mirror near the top end of the tube. Many observers argue that this allows a more comfortable viewing position.

From the point of view of the planetary observer, the main limitation in Newtonian reflectors is the loss of image contrast due to the central obstruction necessarily introduced by the secondary mirror and its holder. If the telescope is to be used as a dedicated planetary instrument, this can be alleviated to some extent by the use of a smaller secondary mirror; this, however, will limit the telescope's usefulness in other areas, such as deep sky observing. Some observers avoid this problem by having interchangeable secondary mirrors for use with their Newtonian reflectors. Care must be taken, of course, to ensure correct alignment (collimation) of the optics in such an instrument to obtain optimal performance.

Aperture for aperture, the consensus is that a 150 mm reflector will give planetary observing performance equivalent to a 114 mm refractor (the refractor has in its favor an unobstructed light path). In terms of cost, however, the reflector may prove cheaper. For observing Mars (and also Saturn), where the apparent disk of the planet is small, the greater resolving power of a still-larger (200–300 mm) aperture reflector will prove advantageous, and more cost- (and space-) effective than the equivalent large refractor – indeed, refractors of such aperture are hardly ever built!

A 150 mm aperture reflector at $f/6$ will just about be as convenient to set up and unmount of an evening as a 114 mm refractor. Larger reflectors might demand a permanent mount, but the optical tube assembly can be kept indoors when not in use.

A consideration with reflectors is the production of air currents in their open-ended tubes, which lead to unstable images early on in an observing session as the telescope cools down. This can be at least partially avoided by setting the telescope up outdoors an hour or two in advance of the planned observation, allowing it to equilibrate in temperature with its surroundings. This is less of a problem, obviously, with telescopes permanently mounted in outdoor observatories. Newtonian reflectors with "skeleton" tube assemblies solve the problem of tube currents by employing trusses rather than a solid unit. These allow free circulation of air, but have the disadvantage, if used in the open, of possibly allowing stray light into the optical path.

▶ A Newtonian reflector is shown here on an equatorial mounting. Light from the main mirror is directed via a secondary into the eyepiece housing near the top of the tube, offering a relatively comfortable viewing position.

Newtonian reflectors are hardly designed for daytime use, and it will be rare indeed to find models adapted for terrestrial viewing, with the addition of a "correcting" lens to produce an erect image.

Catadioptrics

The air currents that produce unsteady views in the open tubes of Newtonian reflectors are not a problem, naturally, for the sealed optics of a refractor. Catadioptric systems, which have become very popular with amateur astronomers since the early 1980s, likewise have the benefit of a sealed unit. Catadioptrics use a combination of a main light-gathering mirror coupled to refracting lenses to "fold" the long focal-length light path into a reasonably short (and portable) tube. In some respects, the Meade and Celestron 200 mm aperture Schmidt–Cassegrain catadioptric telescopes (SCTs) have replaced the 76 mm refractor or 150 mm Newtonian reflector as the "typical" amateur telescope in the past decade or so.

SCTs are versatile instruments, offering plenty of aperture for deep sky observing, while also working well with high-power eyepieces for planetary work. As in a Newtonian reflector, the secondary mirror produces some obstruction with concomitant loss of contrast, but the degradation of the image is not too serious. At the September 1988 perihelic opposition, I enjoyed some very crisp views of Mars through a 200 mm SCT, and many observers find such instruments ideal for visual, photographic and CCD observations. A 200 mm aperture SCT is quite a heavy instrument, and will take a reasonable amount of time and physical effort to set up and take down.

An SCT brings light to a focus through a hole in the main mirror, such that the observer is looking straight through the tube; focusing

▶ Since the 1980s, Schmidt–Cassegrain catadioptric reflectors (SCTs) have become popular, offering large apertures in a compact tube. This instrument is on an equatorial fork mount.

▶ An equatorial mounting allows a telescope to be driven to follow the apparent motion of the stars and planets due to Earth's rotation. The equatorial is based around two main axes. The polar axis, around which the driving motor turns the telescope, is aligned parallel to Earth's axis of rotation. North and south movements are made along the declination axis.

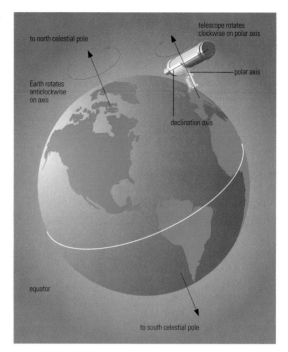

is achieved by small movements of the mirror itself. The secondary mirror is usually a silvered area on the inner surface of the clear front-end correcting plate; it is optically figured to compensate for the main mirror's spherical aberration, and also seals the tube. As with refractors, catadioptric telescopes are commonly fitted for viewing comfort with a right-angle star diagonal, which will again make recognition of Martian features difficult by laterally reversing the view!

The ideal planetary telescope, then, may be an achromatic or – if you can afford it – apochromatic refractor of 80–114 mm aperture, or a 150–300 mm aperture Newtonian reflector, or a 200 mm or greater aperture SCT. Whatever telescope you have, the important thing is to use it, become familiar with it, and enjoy it!

Mountings

A very important consideration for high-magnification viewing of Mars – or any other astronomical object – is the telescope mounting. An unstable mounting, which trembles at the slightest touch of the focuser, or breath of wind, is worse than useless, and will soon lead to disillusionment.

An observer with a small, portable instrument may use a simple altazimuth mounting, allowing vertical (altitude) and horizontal (azimuth) motion. Some small telescopes even have adaptors that allow them to be fitted on standard camera tripods for convenience. If the tripod is strong enough, it is possible to obtain reasonable views with such a set up, provided vibrations quickly subside after the necessary frequent adjustments of the telescope's aiming direction. In the absence of a drive mechanism to counter the stars' apparent motion due to Earth's rotation, the observer will continually have to nudge the telescope westward in order to follow the target object as it drifts through the eyepiece field. At ×200 magnification, for example, an observer will have only a minute or so of viewing time before Mars leaves the field! This is inconvenient but not insurmountable: the determined, patient observer can live with the problem.

A larger telescope will usually be set up on an equatorial mounting, which rotates around an axis pointed at the celestial pole. This motion can be driven by a motor to counter Earth's rotation, and if the mounting is well set up and aligned, objects should remain in the field while the drive is on, allowing the observer to concentrate on the details in the eyepiece, rather than tending to the telescope's aiming direction. For imaging Mars, either photographically or with a CCD or video camera arrangement, a well-aligned, accurately driven equatorial mount is pretty much essential.

Eyepieces

Regardless of how good a telescope's objective lens or primary mirror might be, its performance depends heavily on the eyepiece inserted at the other end of the optical path. There is little point in paying a lot of money for a good-quality telescope, then limiting its performance with cheap, poor-quality eyepieces, which themselves suffer from chromatic aberration.

In the past, a bewildering range of eyepiece types, with differing optical configurations and properties, was available. The range on the market today is in many respects more straightforward, and eyepieces suitable for planetary observation are easy to obtain.

It is quite common for new telescopes to come equipped with Plössl eyepieces. These eyepieces are made up of paired elements, separated by a small air gap. They are perfectly adequate for planetary observing. Less common nowadays are solid glass configurations like the Tolles or Monocentric, which were highly regarded by observers in the past for their lack of ghost images (a result, in some air-spaced eyepiece types, of internal reflections).

Eyepieces are usually described in terms of their internal focal length (inscribed on the barrel or casing). The magnification can be determined by dividing the telescope's focal length by that of the eyepiece. A 9 mm eyepiece on an f/6 (900 mm focal length) 150 mm aperture Newtonian reflector, for example, would give a magnification of ×100; the same eyepiece on an f/8 114 mm refractor (focal length 972 mm) would give a magnification of ×108.

Most observers like to have at their disposal a number of eyepieces at different focal lengths, thereby offering a choice of magnifications. An eyepiece of 4 mm focal length on the respective instruments above, for example, would offer still more highly magnified views, of ×225 and ×243.

Bear in mind, however, that there is a limit to how much magnification your telescope can offer without loss of image quality. Even in a top-notch instrument, high-power views will also be relatively dim. A rule of thumb is that the highest magnification one can expect to deliver good images is roughly ×50 for every 25 mm of telescope aperture, which means about ×300 for a 150 mm instrument, ×228 for a 114 mm, and so on. Some experts suggest that a 150 mm telescope might just take ×350, and a 114 mm × 270, on a good night. On nights when atmospheric turbulence is bad, however, it may not be possible to use high magnifications in any case.

A good way of extending the available range of magnifications is to use a Barlow lens – an adaptor that fits into the focuser and has the optical effect of doubling or trebling the focal length of the telescope. Essentially, purchase of a Barlow can double the range of magnifications available with your existing eyepiece set!

Filters

Many of the details on Mars are rather elusive, showing low contrast under normal viewing conditions. Eyepiece filters can be used to increase the contrast, rendering these features more readily visible.

Filters are normally mounted in small holders which screw into the threaded barrel of most eyepieces on the side closer to the objective (the "field" side as opposed to the "eye" side). They are available in a wide range of colors. The standard set used by experienced planetary observers is the Kodak Wratten series, available from larger telescope and accessory suppliers.

Useful filters for observing Mars are: the yellow-orange Wratten 15 and 21, which enhance detail in the planet's dark surface features; Wratten 38A (blue) and 47 (violet), which reveal white clouds in the Martian atmosphere; and Wratten 56 (green) and 25 (red), which shows up yellowish dust clouds.

Other considerations

Whatever your telescope, as an amateur astronomer interested in observing the sky, your main adversary – apart from the weather – will usually be light pollution. Light pollution is the unwanted direct glare from poorly shielded outdoor lights or windows in the vicinity of the observing site, and the all-pervasive general background glow cast upward into the sky by streetlights and other sources of illumination in urban areas. For many forms of astronomical activity, light pollution is a real problem. Long photographic exposures designed to capture faint nebulae may be spoiled by the orange cast of sodium lighting, while visual observers have little chance of seeing dim targets from light-polluted areas.

To some degree, planetary (and lunar) observations are less hindered by light pollution, and remain accessible to town-based astronomers. Mars at its best is a bright enough object to cut through all but the most severely overlit urban skies, and its disk will still give up at least some of its details telescopically. From the point of view of the eye's sensitivity to subtle contrast differences, however, a dark observing site remains advantageous.

When observing faint objects visually, amateur astronomers depend on physiological changes in the eye leading to what is known as dark adaptation. Under prolonged exposure to low-illumination conditions, the pupil of the eye dilates, enabling it to collect more light. Within the eyeball, formation of the pigment rhodopsin enhances the ability to see faint objects. Most importantly, the retina's rod cells take over as the main light-collectors (as opposed to the color-sensitive cones, which dominate our bright-condition vision). These changes help when it comes to detecting contrast differences in Mars' features at the eyepiece, and ideally the observer should remain dark-adapted as much as possible.

Dark adaptation takes more than twenty minutes to become complete after leaving a brightly lit room, and the observer will wish to retain it as long as necessary. If at all possible, do your observing away from direct lighting, and avoid light pollution as much as you can.

Most amateur astronomers do their recording at night with the help of a red flashlight. Red light has been found to have least effect on night vision, while a strong white-light torch is something to avoid! Bicycle rear lamps can be used, or you may like to try the simple modification that I employ – covering the front-end window of a normal torch with overlapping, light-tight strips of red insulating tape. Making notes and sketches under red light may seem strange at first, but it soon becomes second-nature to the seasoned skywatcher.

A clear view is an obvious, but sometimes overlooked, necessity. For those in the British Isles in 2003, for instance, Mars will be relatively low (about 20–25 degrees above the horizon) in the southern sky, and it will make sense to choose an observing location with a clear view in that direction; local features like trees or the garden hedge may come into play! Observers in the southern hemisphere will encounter similar constraints during the 2007–8 apparition when, for them, Mars will be relatively low in the northern sky when best placed.

Weather is, of course, the main consideration for any sort of astronomical observing. Cloudy skies are the bane of meteor watchers or those trying to capture rare events like eclipses, where the respective action may be confined to a single night, or a window of a few hours. Fortunately, Mars (in common with the other major planets) is around long enough when close to its best for clear skies eventually, surely, to turn up. Unlike the situation for observing meteors or faint deep sky objects, the presence of a bright Moon, even quite nearby in the sky, will not preclude observations.

Coming in late summer for northern hemisphere based observers, Mars' opposition in August 2003 should be favorably timed to coincide with periods of settled weather, bringing plenty of clear nights for observers in Europe and North America. The November 2005 opposition comes closer to a time of unsettled weather during the northern autumn, but it is not uncommon at this time of year to find prolonged spells of calm, hazy, anticyclonic weather, with clear skies and night-time frosts. In 2007–8 observers in the northern hemisphere will see Mars at its best in midwinter, when conditions can either be sparklingly clear and frosty (though this can have its disadvantages, as discussed shortly) or marked by stormy weather. Although showing a rather smaller apparent disk at this time, Mars in 2007–8 will have the advantage, for northern observers, of being high up in the sky, where the air is generally clearer and steadier.

Seeing

The effects of weather and atmosphere go somewhat beyond the simple question of whether it is clear or cloudy. In planetary observing, a more fundamental consideration is *seeing* – the steadiness of the air through which the observer views the target. Seeing should not be confused with transparency, which is the clarity of the air. Indeed, it is ironic that the clearest, sharpest frosty nights, when faint stars and the Milky Way can be seen with ease, are often the worst in terms of seeing.

In the British Isles or the northern United States on a frosty December or January evening, for instance, observers often see

Sirius, brightest of the night sky's stars, twinkling and coruscating through a rainbow of colors in the unsteady air low over the southern horizon as the ground gives up what little heat it has gathered by day. This twinkling of Sirius, and other stars, is produced by the uneven refractive properties of "pockets" of air at slightly different temperatures between the observer and object. The worse the twinkling, the poorer the seeing. On nights when the stars are twinkling vigorously, planetary views in the eyepiece will appear to shimmer with equal vigor, and it will prove difficult (on a really bad night impossible!) to make out much in the way of detail. This, of course, is why high magnifications cannot always be used – the increased magnification simply gives a more obvious impression of just how unsteady the air is between telescope and target!

Seeing varies over the course of the night. Most of the cooling of the ground occurs just after sunset, making conditions particularly turbulent in early evening. By the post-midnight hours, seeing conditions are usually a lot steadier. Some observers rate the last couple of hours before dawn as the best of all for planetary work – but this is of little advantage late in a Martian apparition, when the planet is visible only in the evening sky.

Weather certainly influences the seeing. The recent clearance of a cold front, bringing clear skies, can also be followed by unsteady seeing. Steadier conditions are commonly associated with established, but hazy, high-pressure weather systems (anticyclones). Often the best seeing seems to occur on nights when the atmosphere is less than completely transparent. Autumn evenings when mist or fog is starting to develop are often regarded as bringing good, steady seeing conditions favorable for planetary viewing.

Local geography has a part to play, too. The air above paved or concrete surfaces will often be turbulent, particularly at the end of a hot day, making such sites unsuitable for planetary observing. Views over bodies of standing water or grassland are much steadier, and these factors are worth considering when choosing an observing site.

Regular observers always include an estimate of the seeing in their reports, and there are a number of scales in use. The most widely

THE ANTONIADI SEEING SCALE	
I:	Perfect seeing without a quiver.
II:	Slight undulations; moments of calm lasting several seconds.
III:	Moderate seeing, with larger air tremors.
IV:	Poor seeing, with constant troublesome undulations.
V:	Very bad seeing; even a rough sketch impossible.

employed is that devised by the French-Greek astronomer Eugène M. Antoniadi, a noted planetary observer in the late 19th and early 20th century. The five-point Antoniadi Scale is described in the table.

On most nights, seeing will probably be II to III at a reasonable site. Occasions when seeing is so bad that observations are virtually impossible are relatively uncommon, but so also are those nights when it is rock steady. The observer should learn to make the best possible use, under normal conditions, of those fleeting intervals of a few seconds' duration when the seeing settles down, allowing the best views. The only way to acquire this ability is through practice – experience gained in observing targets like Jupiter or the Moon, views of which are also subject to atmospheric turbulence, will help when it comes to trying to discern Mars' more subtle surface details.

Seeing is always worse close to the horizon, where objects are viewed through a thicker "wedge" of air. In 2003 observers in the northern hemisphere will, by necessity, have to view Mars when it is low in the southern sky, and on many nights the views may not be as spectacular as one might wish, particularly after hot summer days.

Keeping a record

Most amateur astronomers, whatever their particular interests, maintain an observing log, in which they keep sketches, written notes and impressions, and other records of what they have seen. Over time, one's observing log becomes essentially a personal diary, perhaps marked out by specific notable events like auroral storms, nights of strong meteor activity or, of course, memorable telescopic views of Mars!

Keeping a record is worthwhile in a number of ways. It can be instructive, and make you a better, more attentive observer, to go back over past records in order to consider how it might have been possible to make more of a viewing opportunity – we all learn from our mistakes! By making notes as you go along, you will be able to identify which eyepiece consistently gives the best view, or what photographic or CCD exposure to employ for the clearest picture. Equally, a neatly maintained record of your observations may prove useful, in combination with those of other enthusiasts, in following events like dust storms or the seasonal and longer-term variations in the appearance of features on Mars.

Timekeeping for astronomy

In all astronomical record-keeping, it is important to adhere to a standard time system. Use of local time zones makes it harder to compare observations made in different parts of the world, and astronomers have long adopted the system of Universal Time.

Universal Time (abbreviated to UT, or sometimes UTC, the latter standing for Coordinated Universal Time) is equivalent to Greenwich Mean Time (GMT). For observers in the British Isles, GMT is, of course, the time system in operation during the winter months. During summer "daylight saving" (BST in the UK), when the clocks are advanced by an hour, observers have to remember to subtract an hour from civil time when correcting to arrive at UT.

Internationally agreed time zones east of the British Isles are ahead of UT, and observers in central Europe, for example, always have to subtract an hour from civil time, as well as allowing for daylight saving during the summer. In North America, there are several separate time zones across the width of the continent, and observers there will be between five and eight hours behind UT. Conversely, those in Australia and New Zealand will be between eight and twelve hours ahead of UT.

CONVERSION BETWEEN LOCAL TIME AND UNIVERSAL TIME	
Central European Time (CET)	UT + 1 hour
Greenwich Mean Time (GMT)	UT
Eastern Standard Time (EST)	UT − 5 hours
Central Standard Time (CST)	UT − 6 hours
Mountain Standard Time (MST)	UT − 7 hours
Pacific Standard Time (PST)	UT − 8 hours
Alaska	UT − 9 hours
Hawaii	UT − 10 hours
Western Australia	UT + 8 hours
Central Australia	UT + 9.5 hours
Eastern Australia	UT + 10 hours
New Zealand	UT + 12 hours

When recording observations, by all means use the local time in your logbook, but always remember to include the Universal Time, too, for the finished record. For example, an observation of Mars made at 22:24 EST on August 29, 2003 corresponds to August 30 03:24 UT.

As we shall see in chapters 5, 6 and 7 the local time influences which parts of Mars are visible – observers in New Zealand, for instance, will see the opposite hemisphere from observers in Europe on a given date.

Sketching

While many advanced amateur astronomers have taken to CCD imaging as a means of recording detail on Mars and the other planets, for the vast majority of observers sketching remains the method of choice. The trained eye can certainly tease out fine details under conditions of turbulent seeing, when photography is impossible and CCD imagery almost so.

Making drawings of planetary details is at least as much about honesty and accuracy of recording as it is about artistic rendition. Whenever you set out to make a drawing of Mars, do so with an open mind as well as open eyes – do not go to the telescope with a preconception of what you should or should not see, but rather make sure that what you draw is really visible, and not imagined. The visual observer certainly cannot expect to resolve fine detail to a level comparable with the recent Hubble Space Telescope images, and such pictures – commonly published and therefore frequently seen – should not be allowed to influence what you see and sketch.

Before you start

Regular observers make their drawings on ready-prepared "blanks" of standard size. For Mars, this is usually a circle of 50 mm diameter on a black background. One is provided on page 56. Ahead of a night's observing, it is a good idea to prepare several blanks – they can easily be made on a photocopier, for example. On a good, long, clear evening, it might be possible to make two or three drawings, showing the changing appearance of Mars as it rotates.

It may seem unusual, at first, to use such a large blank to draw the relatively tiny disk of Mars as it appears in the telescope. Indeed, most beginners fall into the trap of making their first drawings too small, because what they see in the eyepiece is small. By scaling the picture up, however, it becomes easier to show the relative positions of features, and with practice the observer soon becomes used to drawing on a larger circle.

At the eyepiece

At the telescope take a few minutes to let your eyes adjust to the dark, and gain a first impression of Mars' appearance in the eyepiece on the night, before trying to commit the details to a sketch. Once you are sure you have a reasonable idea of what is visible, start the drawing by marking the outline of the polar cap, if one is visible, then roughly draw in the profiles of the dark markings. At this point, you should note the time (Universal Time) to the nearest minute; this will be needed later when finishing the observation report in your log.

Working reasonably quickly, the next stage in making the sketch is to begin filling in the dark outlines with details of internal variations

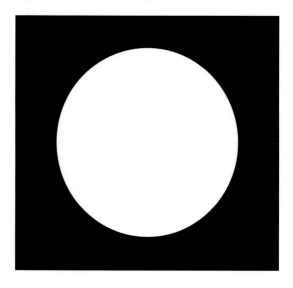

◄ *This Mars drawing blank can be copied and used at the telescope.*

in shading. The sketch should also indicate the position on the disk of any possible light clouds or spots; some observers show the outlines of clouds with a dotted line.

The quality of the sketch is determined, as much as anything, by the seeing. On a steady night, it will be possible to more reliably pick out subtle details. As already discussed, the trained eye becomes attuned to picking out the details in those fleeting instants when the air briefly steadies, and it is during those moments that you should garner the impressions that end up on the sketch.

Alongside the rough sketch in the field, make plenty of notes on the appearance of individual features. These will be useful later on, when you work up your sketch into the final drawing. This is best done after the observing session, when there is more time to ensure the accuracy: the cold and damp of outdoors on an observing evening is hardly the ideal environment for making a work of art! The rough sketch might be annotated with comments such as "bright spot here" (indicated with an arrow) or "hazy patch inside dotted line," and so forth.

Another useful way of describing the relative brightness of Martian features is to make estimates of their intensity on a 10-point scale, taking 0 (the brightest) as equivalent to the pure white of the polar cap(s), and 10 as the darkness of the sky surrounding the planet. Observers may find that the intensity of some of Mars' dark markings varies from one apparition to the next, but it should be borne in mind that such estimates are somewhat subjective.

Sometimes, it may be of interest to concentrate on sketching just one particular area of Mars, and there is nothing wrong with doing so, especially if the observations allow rapid (night-to-night) changes of a feature's appearance to be followed. Fairly simple sketches by amateur observers have been, and still can be, of considerable value in assessing the spread of dust storms, for example, without necessarily demanding a detailed drawing of the entire Martian disk.

For making rough sketches, a soft B or BB pencil is fine. Seasoned observers will usually have a range of pencils at the ready, including one with a sharper point for drawing initial outlines, and others more blunt for drawing in the shading. An eraser for correcting mistakes is another obvious essential! Ideally, the sketch with its accompanying notes should not take more than about half an hour to complete. During this time, Mars will have rotated on its axis by about seven degrees.

As important as making a sketch is remembering to take time on your observing evening to enjoy the view! Once the sketch is completed, there is nothing wrong with simply taking a look at the Red Planet – indeed, many will be content to do so without feeling the need to obtain a permanent record.

After the observation

Tidying up the sketch to produce a finished drawing should, of course, be done as soon as possible after the observation, while the details are still fresh in the mind – preferably no later than the next day.

Soft pencil can be used to render subtle shadings, and the use of erasers or a "blurring" tool, such as an artist's stub (available from specialist craft shops) or even a fingertip, allows production of a final picture similar to what the eye might glimpse in moments of good seeing. A good test of the accuracy is to look at the finished product from a distance of a few meters: does it look the same as the planet did in the eyepiece?

It is, of course, important to note the orientation of the planet as seen in the eyepiece when the drawing was made. An indication of the north and south points on the disk should be given – it does not matter whether the view was inverted (south at the top for observers in Earth's northern hemisphere) so long as this is indicated.

Most observers make their drawings as simple monochrome (black-and-white or grayscale) representations of the planet. Color perception varies from one observer to the next, and it is difficult to quantify the differing shades which each might perceive.

The completed drawing in your logbook, and any copies forwarded to organizations such as the Association of Lunar and Planetary

CHANGES IN THE CENTRAL MERIDIAN LONGITUDE OF MARS

Mars rotates once in 24 hours 37 minutes and 22.6 seconds, or:

1 degree in 4.108325 minutes (4m 06.2s)

15 degrees in 61.557375 minutes (61m 33.4s)

1 minute gives a longitude change of 0.2437 degrees

15 minutes gives a longitude change of 3.6551

60 minutes gives a longitude change of 14.6205

Observers (ALPO) or the British Astronomical Association (BAA) should be accompanied by details of the telescope and magnification used, and of any filters employed during the observation to improve contrast. An estimate of the seeing conditions is another important factor to note.

Recording the time of observation will help the observer (and anyone else making a detailed examination of the drawing) to determine the longitude of the central meridian. In other words, it allows you to work out which areographical longitude was presented toward Earth at the time of observation. (The central meridian is defined as the great circle that connects the north and south points on the visible disk.) Tables allowing computation of the central meridian longitude are published in the annual BAA *Handbook*, Royal Astronomical Society of Canada *Handbook*, the *Astronomical Ephemeris* and elsewhere; some of the more advanced "desktop planetarium" computer programs will calculate the figure for you, while web-accessible ephemerides will also provide the values for 0h UT on a given date. Mars' rotation rate is such that the central meridian longitude increases by 14.6 degrees per hour, and the changing presentation of features is quite evident over timescales of around half an hour.

Knowing the central meridian longitude allows comparison of the finished drawing with existing maps of Mars, and identification of named features. Having a rough idea of which parts of the planet will be on view when can be useful in planning observations: for instance, at a given place on Earth, there will be an interval of about a week during which the prominent dark marking of the Syrtis Major is presented during the evening hours on successive nights.

Photography

Mars' motion over the course of an apparition can be caught on film by taking exposures every few weeks on fast color film (ISO 400) using a standard 50 mm lens at about *f*/2 on a tripod-mounted SLR (single-lens reflex) camera. Exposures kept below 20 seconds' duration will be sufficient to record Mars and the star background down to more or less the naked eye limit without Earth's rotation introducing appreciable trailing. Wide-field shots of Mars among the stars can be attractive, and on occasion it may be nicely framed close to the Pleiades and/or Hyades in Taurus, or in conjunction near – in our line of sight – to one of the other planets. Photo opportunities of this nature are highlighted in chapters 5, 6 and 7.

Photographing Mars through a telescope is, frankly, rather difficult. Much of the reason for this lies with the planet's small apparent size: even at its best in 2003, Mars subtends a diameter of only 25 arcseconds, and it is consequently rather hard to get a satisfactorily large, bright and contrasty image on film.

For wide-field photography of, for example, a star cluster, it is quite common to use the camera at the telescope's prime focus, where the eyepiece would normally be; essentially, the telescope is used as an additional, albeit rather large and expensive, camera lens.

▼ *Wide-field photographs of Mars in conjunction with other planets, or near interesting star groups, can be attractive. This exposure captures Mars (right center) at the end of its perihelic apparition near the much brighter* *Jupiter in March 1989, among the stars of Taurus. The Pleiades are to the right and Hyades to the left. Mars at this time was comparable in brightness and appearance to Aldebaran, the brightest of the Hyades stars.*

For planetary and lunar photography, however, where magnification is paramount, eyepiece projection is the preferred method. The eyepiece is left in place, and used to focus the enlarged image on to the film in the back of the camera. Adapters for eyepiece projection are available from telescope retailers.

Once focused – a tricky operation in some cameras – the planet's image is captured to film in exposures that may require several seconds, thanks to Mars' light being somewhat spread out. A serious shortcoming in photography is the effect of atmospheric seeing, which will degrade image quality over long exposures.

Another major consideration is the film grain. Fast films, which require shorter exposures, tend to have coarser grain, meaning lower resolution. Color films are, in general, grainier than black and white, but many observers prefer the aesthetics of the former. Kodak Technical Pan (TP 2415) is a favorite black-and-white film for applications where fine grain is essential.

Photography on conventional films is certainly a challenge, and for Mars it is best left to those with access to seriously large telescopes (in the 250 mm aperture and upward range). Getting a reasonably large, high-resolution image with smaller instruments is generally regarded as out of the question.

The arrival on the market of ever-cheaper, more sensitive digital cameras makes it likely that "wet" photography, based on darkroom chemistry, will become increasingly restricted to a few specialized areas. Most digital cameras have sufficiently numerous pixels (picture elements) in their light-sensitive recording chips to match the resolution of a reasonably fine-grained conventional film. Digital cameras with good light sensitivity can be used in eyepiece projection for brighter objects like the Moon and planets, and there is the added appeal of being able to review images on the spot, keeping the good ones, while rejecting those marred by intervals of poor seeing and so forth. Digital photography looks to have a bright future for many astronomical applications.

A cheap and popular alternative to digital camera imagery is the use of modified webcams. Stripped out of their computer-top housings, webcams can be adapted for eyepiece projection with acceptable results in planetary and lunar work. Captured to a laptop computer, multiple images can be "stacked" and manipulated using popular processing software, such as Adobe Photoshop or Paintshop Pro. Groups dedicated to the use of webcams for astronomical imaging have sprung up around the world since the late 1990s, and many have websites displaying impressive results and offering guidance and software.

CCD imaging

The single biggest revolution in amateur astronomy in the past 15 years or so has surely been the appearance of relatively affordable charge coupled device (CCD) cameras for imaging work. Of course, the imaging chips in the backs of digital cameras and webcams are from the same family, but CCDs for high-quality imaging offer a greater degree of sensitivity, thanks in part to the attachment of cooling units which reduce the background "noise."

CCDs have been used with some considerable success by amateur observers to record deep sky objects such as galaxies and faint nebulae; by judicious use of dark frames (subtracted from images to remove residual noise) and appropriate processing software, some observers have been able to pull out remarkable images, even from light-polluted suburban observing sites. CCDs have been put to good use in amateur supernova search programs, and in a lot of other areas. In many cases, CCD images now coming from backyard telescopes rival those obtained at professional observatories not so long ago.

Planetary observers, too, have embraced the new technology, and for many of the most dedicated and skilled imagers, like Don Parker (in Florida) or Damian Peach (a British observer now based under clear skies in Tenerife), CCD has supplanted conventional film as the recording medium of choice.

As with film-based photography, the main requirement is to obtain a sufficiently large image on the CCD chip in order to make the most of the telescope's resolving power. So, for CCD recording of Mars and the other planets, the image is again collected via eyepiece projection. The greater light-sensitivity of CCDs relative to film offers the advantage of shorter exposures, thus making poor seeing less of a limitation. However, the observer will require a dedicated computer, housed in the observatory with the telescope, in order to store the image files as they are gathered. Alternatively, a laptop computer may be used for image collection and storage. Image processing can, of course, be done at leisure – the priority

▶ *CCD imaging has enabled amateur astronomers to obtain views of Mars better than the photographic results from professional observatories 20 to 30 years ago. Damian Peach* *recorded this fine view showing Syrtis Major and Hellas at left, Sinus Sabaeus across the middle of the disk, and morning cloud at the following limb on July 2, 2001. South is at the top.*

◀ This webcam image of Mars on July 1, 2001 was obtained using a 510 mm reflector stopped down to a working aperture of 200 mm. A Philips ToUcam Pro was used in conjunction with AstroVideo software developed at the COAA amateur observatory in Portugal's Algarve, where the image was taken by Steve Wainwright. The view is comparable to that shown on page 61, taken a day later.

during the observing run is to obtain a good selection of reasonable-quality raw images.

Monochrome images are generally fine for illustrating the appearance of Mars on a particular night. The main thing to record is the extent and relative brightness of the various surface markings and any clouds that may be present. As with visual recording, it is useful to note somewhere in the finished record the orientation of Mars in the image.

Rather like the computers required to run them, CCDs for amateur astronomy seem to be getting faster and more sophisticated all the time. In the past, for example, color imaging demanded that three separate exposures (filtered in red, green and blue light) be taken, then later combined using processing software to produce a correctly registered picture. "Single shot" color CCDs are now available, though purists suggest that for some applications, including planetary imaging, it is preferable to continue taking the individual filtered exposures. CCD chips for astronomical imaging are now becoming available with greater pixel density, which, by analogy with conventional films, is equivalent to finer "grain."

As with digital photography, CCD imaging has the great advantage that exposures can be taken and reviewed on screen quickly, allowing the observer to discard those degraded by poor seeing. Provided they are taken in quick succession, the best images in a series can be "cherry-picked" and combined in the computer to improve contrast and resolution.

The speed with which images can be recorded in CCD camera systems lends itself to filtering. Use of the same Wratten filters employed by visual observers allows images to be taken in quick succession, to highlight dust clouds, ice clouds or the surface features. In principle, filtering can be applied to conventional photography, but the fainter image of Mars in such cases further extends the length of the exposure, making it likelier that degradation caused by seeing and errors in the telescope motor drive will render the picture less useful. CCDs are far better suited to this work.

Video

Some modern camcorders have sufficient light sensitivity (0.01 to 0.05 lux) for planetary or lunar imaging. With imaging rates of 25 or 30 frames per second, it is possible to use video cameras to obtain clear pictures, getting around the problems introduced by poor seeing. "Frame grabbing" software can be used to select the clearest individual images, which can, in turn, be stacked electronically to enhance the detail. While good results are possible with camcorders, dedicated imaging CCDs will still give the best pictures.

When at its best, Mars is a suitable subject for telescopic observation. As in many other areas of astronomy, however, photography is gradually being edged out by more efficient electronic means, which are less prone to the whims of atmospheric seeing (and certainly a good deal less messy!). Electronic methods also allow the image to be processed using software designed to bring out the best from raw data. It remains a serious caveat, however, that observers should not over-process their raw images – there is only so much information in there, and one should certainly avoid the temptation to stretch and color pictures in order to meet expectations raised by the images published in glossy magazines. Honesty really is the best policy when seeking to image the Red Planet.

While technology marches on, and presents ever more opportunities to rapidly and – within the bounds of the cautions stated above – objectively record the planet, drawing Mars with pencil and paper is still a valid skill, and there will be nights when the trained eye can still make out more than even the best CCD camera, gazing out through Earth's turbulent, murky atmosphere. Visual telescopic observation has served well for something like 350 years, and it is not about to go away!

Above all, the important thing is to make best possible use of good observing opportunities as and when they come along. There is much to be said for the thrill of seeing for oneself some of the details revealed by more sophisticated imaging equipment. With Mars relatively close and certainly prominent in 2003 and 2005, and still well placed for many in 2007–8, the next several years afford the chance for more people to become familiar with this fascinating neighbor world.

THE VIEW FROM EARTH

Of the five naked eye planets known to the ancients, Mars must in many ways have seemed the most perplexing, not only because of its strong red color, but also as a result of its marked variations in brightness. Astronomers describe the brightness of stars and planets in terms of magnitude, a system that dates back to classical Greece and the celestial catalog of Hipparchus of Nicaea. Hipparchus ranked the brightest stars as first magnitude, those slightly less bright as of second magnitude, and so on down to the faintest visible to the naked eye at sixth magnitude. Modern astronomers have taken this relationship and put it on a more rigorous mathematical basis, such that a difference of one whole magnitude is equivalent to a factor of 2.512. In this system, the smaller the numerical magnitude value, the brighter the object – so a magnitude 0 star like Vega in Lyra is a lot brighter than, say, magnitude +3 Albireo in the neighboring constellation of Cygnus.

The very brightest stars and planets have negative values in the modern system (which uses Vega as the "standard"). Sirius, the brightest of all stars in the night sky, has a catalog magnitude −1.5, while Jupiter can be as bright as magnitude −2.8, and Venus magnitude −4.5.

Mars ranges over a factor of about a hundred in brightness. When almost on the far side of the Sun, Mars is around magnitude +2; near the beginning and end of each apparition, the planet is barely

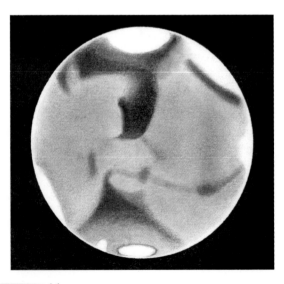

◀ This drawing of Mars was made by R.J. McKim, using a 216 mm reflector ×232 on March 21, 1997; central meridian longitude (CML) = 288°. In all the drawings of Mars in this book, south is at the top, as in the conventional inverted astronomical telescope view from Earth's northern hemisphere. The north polar cap is small, with an outlying fragment called Olympia. Hellas, at top, is bright. Syrtis Major is prominent near the middle.

noticeable. At perihelic oppositions like that of 2003, Mars attains magnitude −2.9, outshone only by the Moon and Venus. Oppositions that coincide with Mars' aphelion see the planet reach a peak brightness of magnitude −1.4.

The principal reason for these variations is, of course, Mars' distance from Earth. When relatively close – as in August 2003 – it presents a larger sunlight-reflecting disk toward us. At perihelic oppositions, the maximum apparent disk diameter on the sky is 25 arcseconds, while at aphelion the opposition apparent diameter is only 14 arcseconds.

When at it farthest from Earth – more than two astronomical units (299 million kilometers/186 million miles) away on the far side of the Sun – Mars has an apparent diameter of only 3 to 4 arcseconds, so it is barely perceptible as a disk in some small telescopes. Indeed, at such times, sensible telescopic observation is pretty much impossible, with Mars appearing no larger in the eyepiece than the similarly unrewarding distant outer gas giants Uranus and Neptune.

The telescopic view

For about half the time, Mars is too far away to show much detail in all but the largest amateur telescopes – reflectors in the 250 to 300 mm aperture class, say – and then only when seeing conditions are exceptionally good. However, in the months close to opposition, even in an unfavorable year when Mars is seen at its best near aphelion, there is an interval of several months during which the apparent disk has a diameter in excess of 10 arcseconds, which is sufficient to allow something as small as an 80 mm refractor to reveal some detail. In the coming years, this "window" is longest, at over 210 days, for the 2003 apparition, shrinking to roughly 135 days for the 2007–8 apparition.

Seen through a typical amateur telescope, Mars is usually fairly close to circular; its small ochre disk (more orange than red when magnified in the eyepiece) has patchy dark markings, and a bright white polar cap is visible by contrast. The Martian atmosphere is full of dust, so if a major, globe-encircling dust storm is in progress, the dark markings below will be rendered invisible and, even in a large telescope, the disk will appear featureless until the storm subsides.

Directions on the visible disk can be described in terms of the planet's orientation relative to the sky. In the course of the night for a naked eye observer anywhere on Earth, the stars – and the Moon and planets – move across the sky from east to west as a result of Earth's rotation. In the northern hemisphere, this motion is left to right, with the object reaching its highest due south in the sky; in the southern hemisphere, motion is right to left, with the highest point being reached due north. For a planetary disk, it is sometimes convenient to describe directions in terms of preceding and following (often abbreviated p. and f.). In an undriven telescope, the preceding limb is that which leaves the field of view first, and is therefore western-most in the sky, while the easternmost limb is described as following.

In most dedicated astronomical telescopes, the image will be inverted relative to the naked eye view. Observers in Earth's northern hemisphere will therefore see Mars with north at the bottom, south at the top; the preceding limb is to the left in the eyepiece view. Features on Mars appear over the following limb, rotating out of view at the preceding limb. At the following limb, it is close to dawn on Mars, while it is evening at the preceding limb.

From Earth's southern hemisphere, an astronomical telescope's inverted image will show Mars' north uppermost in the telescope view. Mars still appears to move west in the undriven field, but toward the right as seen in the eyepiece. Preceding and following limbs are to the right and left respectively, and are equivalent to those seen by northern observers – the only difference is in the north–south orientation (north is at the top in the astronomical telescope view

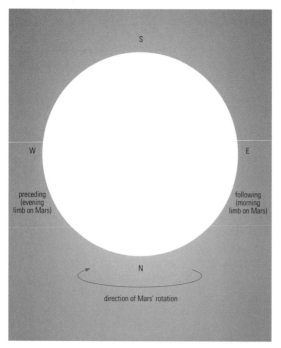

◄ Directions on Mars' visible disk are sometimes described in terms of the planet's apparent motion in the field of view of an undriven telescope. These are summarized here for an observer in Earth's northern hemisphere using an astronomical telescope, giving an inverted (but not laterally reversed) view. South is at the top. West on the sky is to the left, and it is towards this direction that Earth's rotation carries Mars in the course of the observation. The leading – preceding – limb (edge) of Mars in this view marks the side at which features rotate from view.

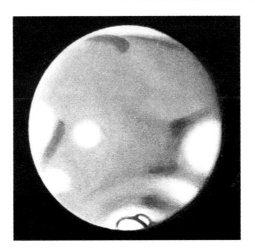

▶ *Mars can at times show a pronounced gibbous phase. This drawing, by R.J. McKim, using the Cambridge University Observatory's 320 mm refractor at ×320, shows bright dawn terminator clouds over Elysium at the following limb, while clouds over the Tharsis volcanoes are seen on the preceding limb. CML = 169°.*

from Australia, New Zealand and South Africa), but the preceding and following limbs remain at west and east relative to the sky respectively, and still mark evening and dawn sides of Mars respectively.

Coordinates on Mars itself are measured in terms of latitude and longitude, with the latter increasing westward. It is usual to describe features at, say, 97°W as being at 097° longitude, while 23°E is presented as 337°. As described in Chapter 3, the visible longitude increases by just over 14 degrees per hour as Mars rotates on its axis.

When close to quadrature, 90 degrees east or west of the Sun in Earth's sky, Mars can show a pronounced gibbous phase, being 84% illuminated (making it similar in appearance to the Moon a couple of days before or after Full). Of the superior planets, only Mars shows such an obvious phase effect – Jupiter and Saturn are too distant at quadrature for the phase to be apparent. The dark segment appears on the evening (preceding) limb on Mars' disk when the planet is emerging into the morning sky after conjunction, and on the dawn (following) limb as Mars moves into Earth's evening sky toward the end of an apparition.

The small disk of Mars does not give up it details easily! Indeed, the first telescopic observers soon moved on to other, more instantly rewarding targets. Early 17th-century optics were rather poor in comparison with those available now, and it is hardly surprising that Galileo and others saw little to invite closer inspection of Mars. Later observers, using better optics, were more able to make out the dark markings.

In keeping with the system adopted for naming features on the Moon, cartographers of Mars in the second half of the 19th centu-

MARS - February to March 2002

Feb 15th   Feb 16th   Mar 1st

17:45 UT   17:33 UT   17:44 UT

CM=228.7   CM=215.9   CM=90.2
D=5.09"   D=5.07"   D=4.82"

D. Peach

◄ *Mars' south polar cap is visible at top in these three CCD images by Damian Peach. They were taken in early 2002, near the end of the 2001–2 apparition.*

ry assigned names like *Mare* ("sea") and *Sinus* ("bay") to the markings on Mars. Some of the observers of this time held to the idea that the changing outlines of certain dark features were caused by shallow Martian seas flooding surrounding marshland to varying degrees. Although we have long known Mars to be arid, it has remained an observational convenience to retain such nomenclature. Many of Mars' extensive featureless areas have been named for terrestrial deserts, with titles such as Arabia, Sinai and so on.

The visible markings on Mars as seen from the distance of Earth bear limited relation to the geological features imaged in close-up from spacecraft. As discussed in Chapter 2, we were unaware of the existence of craters on Mars prior to the 1960s, while even the huge Valles Marineris complex and the true nature of the giant Martian shield volcanoes only became apparent during the Mariner 9 orbiter exploration of 1971–2. The map of Mars' light and dark regions – albedo features (regions that differ in reflectivity) – is quite hard to reconcile, in places, with the surface geology.

Polar caps

In a small telescope, the polar caps are perhaps Mars' most obvious features, by virtue of their high contrast relative to the background. At perihelic oppositions, the larger, south polar cap is presented well toward Earth, while we have a better – but more distant – view of the north polar cap when Mars reaches opposition at aphelion.

The caps grow during autumn and winter in their respective hemispheres, becoming shrouded by mists and fogs of water ice or carbon dioxide clouds. This coverage is initially patchy, and then grows to envelop the whole cap, dispersing again during the spring to expose the surface ice (water and carbon dioxide) deposits. As these deposits melt, the ice caps retreat toward the poles. Spring/summer melting of the southern cap is more rapid than that of the northern, while the northern shrinks to a smaller remnant than does the southern. Neither cap disappears completely. During the spring, the melting cap is surrounded by a dark rim, which also retreats toward the pole, then fades as summer approaches.

Careful observations show that the polar caps do not melt evenly. The springtime retreat of the northern cap is punctuated by fragmentation into a number of distinct areas, separated by narrow dark features named *rimae*. (In standard planetary nomenclature, a rima is a fissure.) One of the most prominent is the Rima Borealis, which isolates a fragment named Olympia (at longitude 200°, latitude 70°N) from the main body of the north polar cap. At midsummer, the largest part of the cap is split in two by the Rima Tenuis.

Observation of north polar cap features is difficult, given that they are only really favorably presented at a time when the apparent Martian

▼ This drawing was made by R.J. McKim using a 360 mm refractor ×450 (Arcetri Observatory, Florence) on August 17, 1988; CML =261°. Novus Mons (the Mountains of Mitchell) can be seen detached from the main body of the south polar cap. Hellas, Syrtis Major and Mare Tyrrhenum are well visible.

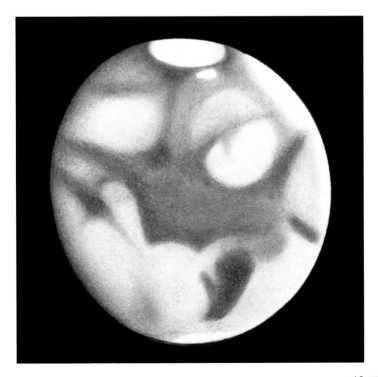

disk is rather small. The study of these features, therefore, tends to lie with experienced observers using large telescopes. Excellent work in this respect was carried out by members of the Association of Lunar and Planetary Observers (ALPO) during the 1980s.

Retreat of the south polar cap is a little easier to follow in more modest telescopes, particularly at perihelic oppositions, such as in 2003, when closest approach coincides with summer in Mars' southern hemisphere. Thanks to the eccentricity of Mars' orbit, summertime heating in the southern hemisphere is greater than in the northern, and the shrinkage of the southern polar cap is more rapid than that of the northern. As in the north, spring and summer melting leaves behind isolated regions of ice as the cap retreats poleward. Most prominent of these are the Mountains of Mitchell (named for Ormsby MacKnight Mitchell, of Cincinnati Observatory, who first reported the feature in 1845), more formally known as Novus Mons, at latitude 75°S, longitude 320°.

While their retreat in summer is observable, the re-growth of both the polar caps is hidden beneath the autumn mist shroud.

Albedo features

For the most part, Mars' telescopic disk appears dominated by an orange-red pale background on which grayish-green dark markings are superimposed. It is hardly surprising, from the color of Mars'

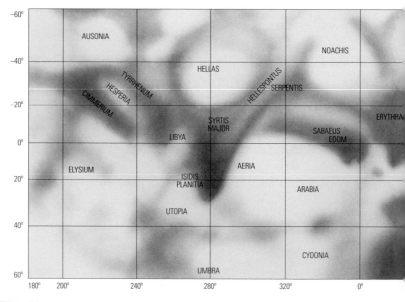

darker areas, that some 19th- and early 20th-century astronomers presumed these to be the result of simple plant growth or lichens. Support was even provided for this notion by the seasonal "wave of darkening," during which the markings were found to become more pronounced with the advance of spring toward summer in the respective Martian hemispheres. The 1976 Viking landers found no evidence for plant life in the Martian soil, and the idea that the dark markings represented patches of growth had long been discounted. There is strong evidence to suggest that the wave of darkening is in reality caused by lightening of the ochre background: contrast between the background and the dark features is enhanced as winter deposits of dust are uncovered.

Maps of Mars' albedo features have been compiled from centuries of observation, and while it has been changed somewhat over the years, a standard nomenclature for the visible markings is now in use. An arbitrary longitude zero on Mars was defined by the Sinus Meridiani (later renamed Meridiani Terra), a relatively obscure feature just south of the planet's equator. Longitude increases westward from the standard meridian.

The most prominent of all Mars' dark features is the V-shaped Syrtis Major. Centered around longitude 290°, it straddles the equator, with its sharper vertex extending to about 30° latitude in the northern hemisphere.

◀ Mars' albedo features are marked on this map. South is at the top.

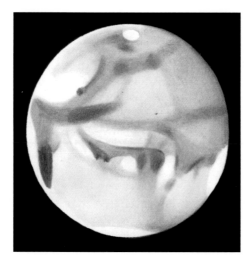

◀ This drawing was made by R.J. McKim using the Meudon Observatory 830 mm refractor ×400, on October 26, 1988; CML = 350°. The small summer south polar cap is seen, with Hellespontus, Syrtis Major, Sinus Sabaeus and Sinus Meridiani.

Syrtis Major shows marked variations in shape over the course of the Martian year. At midwinter in the northern hemisphere – when Mars is close to perihelion (opposition in 2003, for instance) – Syrtis Major is narrower and more tapered than it is at midsummer. On a good night, subtle structure can be seen in the feature's shading. Syrtis Major has shown changes in overall shape on timescales of decades, sometimes appearing broader and more blunted. These changes can be put down to shifting coverage of underlying dark material by wind-whipped dust. Geologically, Syrtis Major is a plateau in the rolling plains that dominate Mars' northern hemisphere.

The region just east and north of Syrtis Major – Thoth-Nepenthes – has undergone more dramatic long-term changes. Linear dark markings that used to be conspicuous in this area are now barely distinguishable from the general reddish background of the Martian deserts.

Due south of the Syrtis Major, at 44°S latitude, and best presented when Mars' south pole is tilted toward Earth, is the huge, more or less circular Hellas. Revealed by spacecraft as an impact basin, Hellas appears light by contrast with the neighboring Syrtis Major. It shows variations in brightness depending on whether frost deposits or fog are present on its deep floor. Obscuration by dust can also occur in this region, as will be discussed later.

Extending westward from the broader end of Syrtis Major for about 60 degrees in longitude, roughly along the 10°S parallel, is Sinus Sabaeus, another of Mars' more prominent dark features. Like Syrtis Major, this feature has shown some variation in shape over long timescales. The western end of Sinus Sabaeus ends in two northward-pointing "prongs." These mark the Meridiani Terra (commonly called Dawes' Forked Bay by observers in the past), which is used as the arbitrary longitude zero. The name Meridiani Terra was officially dropped from maps by the US Geological Survey in 2001, but the feature remains reasonably prominent in modest-aperture telescopes under good conditions.

Immediately west of Hellas and south of Sinus Sabaeus is Mare Serpentis – another dark albedo feature prone to dust obscuration.

West from Sinus Sabaeus is the Mare Erythraeum. It is mainly in the southern hemisphere, although a few elongated patches stretch

▼ *The Argyre impact basin is bright on the south limb, bordered by Mare Erythraeum, in this drawing by R.J. McKim made on March 10, 1997, using a 216 mm reflector ×464. Coprates, Aurorae Sinus and Solis Planum are visible on the south following side of the disk. The large, dark feature in Mars' northern hemisphere is Mare Acidalium, with Chryse to its south. CML = 021°.*

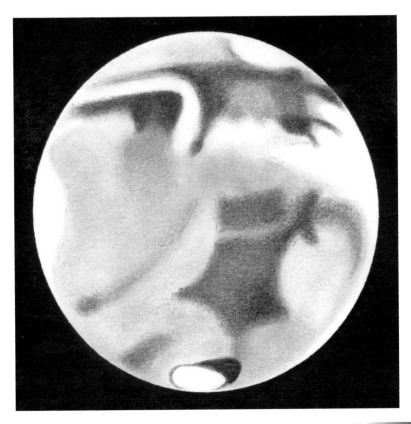

just north of the Martian equator. Another dark feature, it is less prominent than Syrtis Major. Mare Erythraeum, and the adjacent Coprates region, is the site of the Valles Marineris canyon complex, which was revealed by spacecraft exploration. The Valles Marineris itself remains beyond reach of Earthbound telescopes.

South of Mare Erythraeum is the second of Mars' two great impact basins, Argyre. At 50°S, 40°W and about half the size of Hellas, Argyre is best seen at perihelic oppositions when Mars' south pole is tilted toward us. Like Hellas, Argyre is prone to subtle night-to-night changes in appearance caused by clouds and haze.

The northern hemisphere around this longitude (centered roughly at 30°W) is dominated by the huge, fairly dark Mare Acidalium. Another vast plateau area, it is identified on geological maps as Acidalia Planitia.

▼ *This fine view of Solis Planum, also known as "The Eye of Mars," was drawn by R.J. McKim on August 27, 1988, using the 360 mm refractor at Arcetri Observatory, Florence, and a magnification of ×450. CML = 110°.*

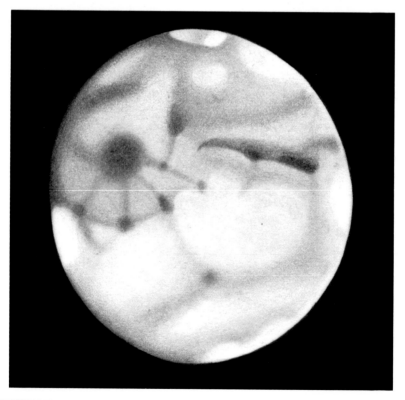

JANUARY 2nd, 2003

07:02 UT
CML=177.5

Diam=4.59°

► *The side of Mars centered on the Amazonis desert and Elysium is, perhaps unkindly, described by some observers as "less interesting." A better description might be "more challenging," and in this CCD image from early in the 2003 apparition, Damian Peach has succeeded in recording much subtle detail.*

Between Mare Acidalium in the north and Mare Erythraeum in the south lies a lighter area, which is mottled with smaller dark markings. This corresponds with the Chryse Planitia region in which the Viking 1 (1976) and Mars Pathfinder (1997) landers came down. The area is dominated in close-up spacecraft images by water outflow features (again, none of this is evident in our more remote view from Earth).

West from Mare Erythraeum, and centered around longitude 87° and latitude 25°S, is Solis Planum, which is one of Mars' more intriguing albedo features. Formerly known as Solis Lacus ("Lake of the Sun"), this feature has long had the nickname of "The Eye of Mars." An oval feature, Solis Planum is somewhat darker than the Syria and Thaumasia deserts that surround it. At its center is a darker spot, which lends to the "Eye" illusion. Dark streaks, which run from the center to the periphery, often show changes in configuration. Solis Planum can vary considerably in extent and appearance from one apparition to the next thanks to rapid shifts of dust deposits in the region.

Northwest of Solis Planum is the Tharsis region, which is dominated by the giant volcanoes imaged by Mariner 9 and subsequent orbiting spacecraft. Clouds forming over the elevated Tharsis ridge are visible from Earth, but claims of visual detection of the volcanoes themselves are mostly treated with scepticism. Clouds may reveal the position of Olympus Mons to the west of Tharsis.

The region from Tharsis westward through the vast Amazonis desert and on to Elysium in the southern hemisphere (longitude 90° to 240°) is generally quite light, with only very subtle dark markings. Some observers comment wryly that Mars has two sides – an interesting one, where Syrtis Major, Hellas and Solis Planum are found, and a rather bland, less interesting one, which is dominated by largely featureless desert.

White clouds

Although extremely dry by terrestrial standards, the Martian atmosphere can sustain formation of clouds and hazes, and these can occasionally be seen through Earth-based telescopes. Use of a Wratten 38A (blue) filter helps to improve contrast between the clouds and the reddish surface of the planet.

White clouds are commonly seen in association with the polar caps (the autumn–winter polar hood), but they also occur at other sites. Melting and evaporation of polar ice deposits in spring and summer increases the amount of water vapor available for cloud formation. Therefore, in a given hemisphere of Mars, white clouds will be commonest between spring and autumn when water vapor is more plentiful, but rare in winter when water is locked up in the polar deposits.

Some of the observed white clouds appear to be surface fogs in low-lying areas, such as the depths of the Hellas basin or the Valles Marineris region (the Coprates albedo feature). Distinctive W-shaped white clouds are seen in the Tharsis region, and are associated with its high-elevation volcanic terrain. These clouds are often described as orographic clouds, and they result from air being forced to higher elevations and lower temperatures, thereby reaching the dew point at which clouds will condense. The Tharsis clouds appear to be an afternoon phenomenon, and so are most commonly seen when this region is presented on the preceding side of the planet's visible disk. W-shaped clouds are also associated with Olympus Mons. The feature's previous name – Nix Olympica, meaning "Snows of Olympus" – comes from its appearance to earlier observers, who interpreted the clouds enshrouding the peak as a white spot on the mountain.

In larger telescopes, white clouds can be seen near the morning or evening terminator close to the time of quadrature. Clouds in the high atmosphere are illuminated by the Sun at times when the underlying surface is still in darkness, and so they are seen to protrude beyond the terminator.

Some of the observable white clouds may actually be surface frost deposits. Frost is thought to form frequently on the floor of the Hellas basin, thus accounting for many of the albedo changes seen in the feature.

An interesting phenomenon involving surface frost and ice deposits, or perhaps ice crystals suspended in the Martian atmosphere, is the possibility of specular reflection. Much as a distant window might dazzlingly reflect sunlight back to the observer for a few minutes when the Sun–window–observer angles are just right,

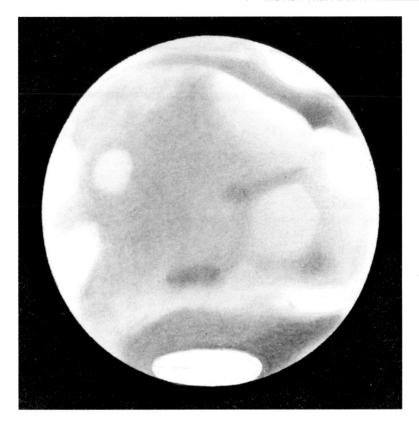

▲ Orographic cloud can be seen on Mars' evening side in this drawing by R.J. McKim, made on February 1, 1995 using a 216 mm reflector ×232. CML = 192°.

so ice and frost on the Martian surface may produce "flares" lasting an hour or so, during which features become unusually bright. Such specular reflection flashes had been suspected in past apparitions, notably in the region of Edom Promontorium, just east of Dawes' Forked Bay in the area of the Sinus Sabaeus. Observing with a 150 mm Newtonian reflector from Florida, Gary Seronik and Rick Fienberg saw just such a brightening in Edom on June 7, 2001. Over the course of an hour, the bright patch showed subtle variations, sometimes fading, then recovering in prominence. Video records were made at the eyepiece of a 300 mm aperture telescope at the same time, and the reality of the phenomenon seems certain. Similar flashes have been reported in the region of Tithonius Lacus, northeast of the Solis Planum.

Blue clearings

Observed through a blue filter (typically Wratten 38A), Mars' surface albedo features show little contrast – usually, only a uniform background and the polar cap will be visible. This poor view is thought to be caused by the presence of a scattering layer high in the Martian atmosphere. Sometimes, however, this layer disperses, affording a clear view through to the albedo features, similar to that in an unfiltered eyepiece.

The phenomenon of blue clearing remains poorly understood, but it may last for several days. Observers can check for its occurrence using either Wratten 38A or 80A filters. Even when a blue clearing occurs, the scattering layer normally remains opaque at shorter wavelengths, into the violet. Sometimes, however, the clearing can be so complete that even in violet filters – which normally reveal no surface markings whatsoever – the albedo features are revealed.

Dust storms

The small particles that make up the Martian soil are easily picked up and transported by near-surface winds, and the constantly shifting pattern of deposition is, as we have seen, a principal cause of the changes in outline and appearance of the planet's dark albedo features, such as Solis Planum and Syrtis Major. Movements of the

▼ These drawings by R.J. McKim show the spread of the 2001 dust storm (yellow clouds), as seen in a 409 mm Dall-Kirkham Cassegrain telescope ×409. At left (July 3, 2001; CML = 316°), Hellas is prominent, and full of yellow dust, while more diffuse dust affects the preceding limb. Syrtis Major remains reasonably clear. By August 10 (CML = 302°), the dust has obscured the albedo features, leaving Hellas and the north polar hood the most obvious visible features.

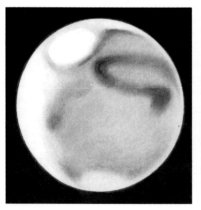
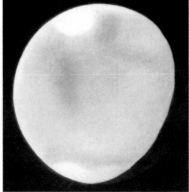

Martian dust are sometimes seen by terrestrial observers as yellow clouds of varying extent.

Yellow clouds covering limited areas can be regarded as relatively local dust storms. On occasion, however, the dust storm may grow to completely envelop the southern hemisphere (as happened in 2001) or, more rarely, the entire planet (1971).

Over the years, amateur observers have amassed a great deal of information on how Martian dust storms develop, and one of the most detailed studies ever made of the phenomenon is that by Dr. Richard McKim, Director of the BAA Mars Section, who has collated detailed archival reports of yellow clouds from the 1890s onward.

Yellow cloud activity is most often seen around Martian perihelion, which is coincident with midsummer in the southern hemisphere and greater warming (insolation) in, for example, the great Hellas basin.

Localized dust storms are quite common in Hellas, and in the area around Solis Planum and Thaumasia. Outbreaks are also found in Argyre and in the Chryse Planitia. Yellow clouds may remain localized or may spread to become regional events that obscure larger parts of the surface. Experienced observers, familiar with the normal appearance of the albedo features, may more readily detect the onset of dust activity. The value of spending many sessions at the eyepiece becoming acquainted with Mars' appearance is obvious for such studies. Use of a Wratten 21 (orange) or 25 (red) filter emphasizes the contrast of yellow clouds.

Globe-encircling dust storms have been recorded in 1909, 1924, 1956, 1971, 1973, 1977, 1982 and 2001. The dust storm at the perihelion of 1971 was notable for it global extent, obscuring features in both hemispheres of Mars from mid-October until December. Although this should have been a very favorable apparition for telescopic observation, in the weeks following the August 10 opposition Mars' albedo features were frequently obscured by first local and then regional yellow clouds, before finally the global storm spread out from Hellas in mid-September. The dust storm encircled the planet in just over three weeks and did not clear completely until late February of 1972. In the history of telescopic observation, no greater Martian dust storm has been seen.

In contrast with 1971, the next perihelic opposition, in 1988, was marked by only local and regional dust storms, allowing reasonably clear views of the albedo features. Writing in the UK magazine *Astronomy Now*, Richard McKim suggests that the 2003 apparition should see some dust activity, but, if long-term trends are followed, perhaps not a major globe-encircling event.

The satellites

Unlike Jupiter's four bright Galilean satellites, or Titan and several others of the moons of Saturn, which are readily visible in small telescopes, Mars' companions, Phobos and Deimos, are extremely difficult to see. At typical respective magnitudes of $+11.3$ and $+12.4$, they are very much fainter than the planet, and especially when Mars is at opposition close to perihelion, they will be swamped by the glare. The angular separation between Phobos and Mars when seen at perihelion is about 35 arcseconds, while Deimos can be found up to 86 arcseconds away; at aphelion, the respective distances are 19 and 48 arcseconds. Advanced observers sometimes aid detection by using an occulting bar, a vertical obscuring strip inserted in the eyepiece to hide the bright planet.

Mars in 2001

Mars was prominent in the skies of May to July 2001, coming to opposition on June 13 as a magnitude -2.35 object among the stars of Ophiuchus (a non-zodiacal constellation through which the ecliptic passes). Although low over the southern horizon for observers in the UK and North America, Mars was still the focus for a lot of attention. The disk diameter of 20.5 arcseconds was favorably large.

At this apparition, Mars was presented with its equator toward Earth, so that neither pole was readily visible. Early in the apparition, shrinkage of the northern polar cap toward the middle of that hemisphere's summer was followed by observers using large telescopes.

White cloud activity was quite frequently noted at equatorial latitudes, particularly in regions such as Chryse. For the most part, the dark albedo features appeared much as they had in several previous apparitions, with Syrtis Major, as usual, the most prominent of the dark markings.

About a week after opposition, yellow cloud activity became apparent in Hellas. Occurring at the beginning of spring in the southern hemisphere, the onset of yellow cloud activity was unexpectedly early in the season. By late June, the dust had spilled out from Hellas to reach Syrtis Major. Observers examining Mars in smaller instruments – I was among them! – were somewhat dismayed to find Mars at this time a rather bland, featureless orb in the eyepiece.

Further dust storms began to appear in the region close to Solis Planum, spreading to merge with that from Hellas during the first week of July. By July 11, the dust storm had become globe-encircling, obscuring the albedo features across Mars' southern hemisphere. From the initial outbreak, it took just over a fortnight for the dust to spread right around Mars!

▶ This fine CCD image of Mars, obtained by Damian Peach on December 22, 2002, offers a preview of the 2003 perihelic apparition. Hellas, at the top (south), appears bright and hazy, while Syrtis Major, on the center of the disk, is dark, broad and prominent.

Views of Mars from Earth were rather unexciting during the rest of July. Some observers with larger telescopes did manage to detect the position of Olympus Mons as a darker spot, protruding above the dust shroud, but overall the view was simply of a featureless yellow-orange veil.

The storm began to subside in late August, but by this time Mars was becoming a more distant target for small telescopes. By October, when the albedo features became more readily apparent, only those with really large instruments could see them well. Localized dust storms continued in the Hellas and Argyre basins. Observers reported no major changes in the outline of Syrtis Major, but Solis Planum had diminished in extent.

As Mars comes to more favorable presentation once again in 2003, observers will be looking carefully for any possible changes following the major dust storm – the biggest in over a decade – during the previous apparition. The albedo features change from one apparition to the next, and over still longer timescales. The circumstances of each opposition can be forecast reliably, but it is less straightforward to predict just how the individual features will appear: Mars is a dynamic world which rewards the patient, attentive observer with changing views in the medium and long term.

MARS IN 2003

Mars was a prominent object in the southern sky during June and July of 2001, shining at magnitude −2 among the stars of Ophiuchus and Sagittarius. The outbreak of a dust storm enshrouding the planet's southern hemisphere limited telescopic views, but to the naked eye Mars was an impressive sight, far outshining its first-magnitude "rival" Antares in the constellation of Scorpius to the west. Through the closing months of 2001 and into 2002, Mars remained visible in the evening sky, almost keeping pace east of the Sun, setting late in the night as it became gradually fainter and more distant. Finally, on August 10, 2002, Mars reached conjunction, lost from view on the far side of the Sun, against the stars of Leo.

The excellent 2003 apparition, with Mars close to perihelion at opposition, got off to a slow start in the closing months of 2002. Now west of the Sun, Mars rose just a couple of hours before daybreak, appearing relatively inconspicuous, at around magnitude +1.5, which is not much brighter than the stars of the Big Dipper, say. Just before Christmas 2002, Mars and Venus (the latter very much brighter, at magnitude −4.5) were fairly close together in the predawn sky.

The real action in the 2003 apparition begins around March, as Earth starts to close in toward Mars, and the planet appears both brighter to the naked eye and larger in apparent telescopic diameter. In early March, Mars is a first-magnitude object among the stars of Sagittarius. It has an apparent disk of 5 arcseconds, which is still too small to reveal much detail, except under remarkably good viewing conditions with a very large telescope. At this time, the planet is rising three hours before the Sun for observers at mid-northern latitudes.

By the end of April, Earth has closed to within one astronomical unit (AU), and the Red Planet is becoming more noticeable: at magnitude 0, it is comparable to bright stars like Arcturus in Boötes or Vega in Lyra. Early May sees the disk diameter pass 10 arcseconds, and at this stage serious amateur observers using large telescopes (250 mm aperture and upward) are able to start more detailed examination of the planet.

In mid-May, Mars can be used by binocular observers as a guide for finding the 8th-magnitude outer gas giant Neptune. On May 13, the pair are separated by not much more than a couple of degrees on the sky, among the stars of western Capricornus.

Mars steadily brightens as it moves eastward now, reaching magnitude −1 in mid-June. Rising not long after midnight, Mars is the most prominent object, apart from the Moon, in early morning skies at this time. The apparent disk diameter swells to 15 arcseconds in late June, at which time Mars passes Uranus – at magnitude +6 an easier binocular target than Neptune – as it moves into Aquarius. Mars spends the next few months in this constellation as Earth finally catches up, slowing Mars' apparent motion against the stars (see the map on page 86).

By mid-July, Mars becomes really prominent at magnitude −2, rising in late evening and dominating the eastern sky by midnight. Its disk has a diameter of 20 arcseconds, and even modest telescopes (100–150 mm aperture) now show some dusky markings and the bright south polar cap.

Over the closing days of July, Mars' eastward motion against the star background slows, then grinds to a halt. From August 1 Mars appears to move westward – retrograde – as Earth speeds by on its faster, inner orbital track (see the diagram on page 21).

Mars rises ever earlier as August progresses, and reaches opposition on August 28. At this time, it lies 0.373 AU (55.8 million kilometers/34.7 million miles) from Earth, which is closer than it has ever been in recorded history.

At opposition, Mars has a large apparent disk diameter of 25.1 arcseconds, and even small telescopes reveal some details. Mars is very bright: at magnitude −2.88, it is the brightest object in the night skies of late August 2003 (the Moon is New, close to the Sun, while Jupiter and Venus are both close to conjunction and lost in the solar glare). Mars is midway between the inverted "Y" asterism of the Water Jar in Aquarius and the first-magnitude star Fomalhaut in Piscis Austrinus.

Perihelic oppositions always occur with Mars well south of the celestial equator. For observers in the southern United Kingdom (latitude 51–52°N), Mars reaches its highest in the sky (culminates) at about 25 degrees above the south horizon at midnight. Observers in central Scotland (56°N) see Mars culminate at a relatively lowly 20 degrees. Further south, however, the Red Planet is better presented. Observers at the latitude of New York (40°N), for instance, see Mars culminate at a more favorable 35 degrees (as do those in Spain; observers in Europe might do well to consider a late-August trip to the Mediterranean in 2003!). From Florida at 25°N, Mars culminates at a very useful 50 degrees' altitude. Those in the southern hemisphere will, of course, see Mars a good deal higher up in their skies, culminating at midnight. At 30°S, for example, Mars culminates at a most favorable 75 degrees altitude above the northern horizon close to opposition.

Around the world, Mars will be noticed by anyone outdoors on a clear August evening as a red beacon in the sky, and public interest is bound to be high at the time. In the United Kingdom, for example, local astronomical societies are organizing a National Astronomy Week to coincide with Mars' prime visibility period at the end of August.

Many will be opening their observatories to the public, offering the opportunity for those who have not got their own telescopes to enjoy live close-up views. Similar events worldwide will be advertised on the Internet, or in astronomy magazines and the local press, and it is well worth looking out for such events in your neighborhood!

Mars rises at sunset when at opposition, and in the weeks just afterward is prominent, as a result of its red color and its brightness, low in the eastern evening sky as dusk falls. Mars remains prominent in the evening skies throughout September and October 2003, staying brighter than magnitude −2 and showing a greater than 20 arcsecond disk until October 5. Retrograde motion stops on October 1, and Mars begins to move steadily eastward against the stars again.

As Earth pulls away, Mars appears progressively fainter and smaller, with its apparent diameter falling below 15 arcseconds on October 25, and its brightness dropping below magnitude −1 on November 7. Early December sees the disk diameter down to 10 arcseconds, and by this time Mars is really only well seen in early evening. By Christmas 2003, Mars is fainter than magnitude 0, and more than one astronomical unit away. Quadrature, 90 degrees east of the Sun, is reached on December 29. Mars then shows a gibbous phase, with the dawn terminator on the following side of the visible disk. At this time, those equipped with large telescopes may see white clouds, or hazes lifted by the early morning Sun, as bright patches close to the terminator.

For most observers, the apparition is essentially over by February 2004, when Mars' apparent diameter falls below 5 arcseconds. The planet's steady eastward progress across the sky from Pisces in December to Aries in February means that for much of the interval Mars sets around midnight. In the third week of March 2004, Mars passes close to the Pleiades and Hyades in Taurus. At this time it is a red magnitude +1 spark to compare with the similarly bright star Aldebaran, which marks the eye of the Bull. On the evening of March 26–27, European observers see the waxing crescent Moon very close to Mars; in parts of northern Canada, Mars is occulted – covered – by the Moon around 00h 30m Universal Time on the same night. Unlike occultations of stars by the Moon, which are more or less instantaneous, those involving the planets (which show extended disks) take several seconds to become complete.

By April 2004, Mars is more than two astronomical units distant, and too small to show much detail. During May, it briefly shares the evening sky with brilliant Venus, but the two never get particularly close in line of sight. Mars is in Gemini in mid-May, close to yellowish, magnitude 0 Saturn; on May 23, the Moon will be nearby, making an attractive grouping for wide-field photography.

Although the apparition seems to take forever to come to a close, Mars is eventually "caught" by the Sun, reaching conjunction on September 16, 2004.

Highlights for telescopic observers in 2003

During the perihelic apparition of 2003, Mars will be presented, with its southern hemisphere tilted toward Earth. At the time of opposition at the end of August, it will be late spring in Mars' southern hemisphere, and during the best parts of the apparition, observers will see the south polar cap shrink and retreat to higher latitudes. The springtime dark "collar" around the cap should remain visible around opposition. By mid-August, the Novus Mons ("Mountains of Mitchell") will probably have become detached from the main cap, breaking up into smaller remnants by opposition at the month's end.

Both the time of night and the date govern which albedo features are best presented for observation. Because Mars has a longer rotation period than Earth, the same feature will come back to the central meridian as viewed by an observer at a fixed location on Earth some 37 minutes later from one night to the next. Over the course of a week or so, there will be a series of opportunities to view a given region of Mars – first in the early evening, then progressively later until eventually the region will not have rotated on to the visible hemisphere before the end of the night.

The slower rotation of Mars also means that the parts of the planet best presented for observation by those in Europe will differ from those best seen in America at that time. For instance, if the Syrtis Major is on the central meridian for observers in the UK at midnight Universal Time on a given night, it will have rotated a long way toward the preceding limb by the time Mars is well on view from the eastern United States some five hours later. Observers in New Zealand and Australia have an even more different view, seeing essentially the opposite hemisphere of Mars from their European counterparts.

Observing Mars at the same time each night over several successive nights will show features ever further east on the central meridian: some describe this as an illusory "backward rotation." Essentially, for features at a given longitude on Mars, there is a favorable viewing interval of about a week once every 40 days or so.

In 2003 observers in the United Kingdom will see the Syrtis Major/Hellas region near the central meridian of Mars' visible disk on evenings during the second week of July, and again in the week leading up to opposition in late August. This part of Mars will again be well presented for those in western Europe during the first week of October. For those in the eastern United States, the best viewing conditions for the Syrtis Major are in the third week of July, last week of August and second week of October. In the western United States, this part of Mars is better seen in late July, early September and the third week of October. As seen from Australia and New Zealand, Syrtis Major will be presented toward Earth on evenings during the opening week of July, early August, third week of September and late October.

The best viewing periods for the "Eye of Mars," Solis Planum, will come in early July, the first week of August and the second week of September for western European observers. For those in the eastern and western United States, it will be presented a week and a fortnight later, respectively. In Australia and New Zealand, observers will find Solis Planum at its best evening presentation in the second week of July, third week of August, and the first weeks of October and November.

The more challenging, relatively featureless side of Mars is presented about a week after Solis Planum is best seen. At this time, observers should look out for white W-shaped orographic clouds forming around the Tharsis volcanic peaks in the preceding (afternoon) half of the disk.

Throughout the 2003 apparition, observers will be on the lookout for yellow clouds indicative of dust storm activity, which is common at the time of Mars' perihelion. Monitoring the occurrence and

▼ *The apparent path of Mars against the star background in 2003–4. Larger* circles indicate greater brightness, as in the magnitude scale at bottom.

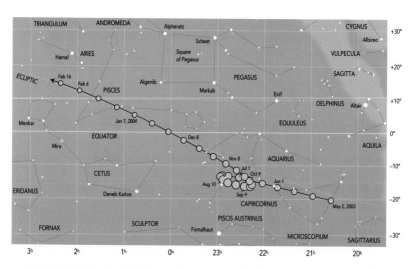

spread of such storms is a truly international effort, requiring cooperation between observing groups in Europe, America and Australia/New Zealand. As a result of Mars' 24 hours 37 minutes rotation, a dust storm first seen in Europe may have to be followed ten days later solely by observers in North America. Equally, observers in Europe will have to wait up to perhaps ten days to see a dust storm first caught from Australia.

The close apparition in 2003 should allow many observers good viewing of Mars' albedo features and south polar cap, barring the onset of a major globe-encircling dust storm.

Mars 2003 apparition timetable

This timetable describes key moments in the apparition. To give a full picture, it begins and ends with Mars at conjunction.

August 10, 2002 Mars is at conjunction with the Sun.

September 2002 to January 2003 Mars emerges gradually into the early morning sky, rising a couple of hours before the Sun. It is far from conspicuous, at magnitude +1.8 to +1.5, as it moves against the stellar background of Virgo and Libra.

January 7, 2003 Mars' distance falls below 2.0 AU (299,200,000 km/185,920,000 miles) from Earth.

January 22, 2003 Mars' apparent disk diameter reaches 5 arcseconds.

February 25, 2003 Mars brightens to magnitude +1.0.

April 17, 2003 Mars reaches quadrature, 90 degrees west of the Sun. It has a distinctly gibbous phase.

April 29, 2003 Mars is now 1.0 AU (149,600,000 km/92,960,000 miles) from Earth.

April 30, 2003 Mars brightens to magnitude 0.0.

May 7, 2003 Mars' apparent disk diameter reaches 10 arcseconds; larger amateur telescopes should by now show some detail.

May 13, 2003 Mars is 2 degrees from Neptune, in Capricornus.

June 15, 2003 Mars brightens to magnitude −1.0.

June 20, 2003 Mars' apparent disk diameter reaches 15 arcseconds; even modest amateur telescopes now reveal some detail. Mars is just over 3 degrees from Uranus in western Aquarius.

July 12, 2003 Mars' distance from Earth falls below 0.5 AU (74,800,000 km/46,480,000 miles).

July 17, 2003 Mars' apparent disk diameter reaches 20 arcseconds.

July 20, 2003 Mars brightens to magnitude −2.0.

August 1, 2003 Mars' retrograde (westward) motion relative to the star background begins.

August 24–30, 2003 Mars' apparent disk diameter reaches 25.1 arcseconds – the maximum for this apparition.

August 26–29, 2003 Mars peaks in brightness, at magnitude −2.88.

August 28, 2003 Mars is at opposition, 0.373 AU (55,800,000 km/ 34,700,000 miles) from Earth. It is visible throughout the night.

October 1, 2003 Retrograde motion ends; Mars begins direct (eastward) motion relative to the star background once more.

October 5, 2003 Mars' apparent disk diameter falls below 20 arcseconds; plenty of detail is still visible in moderate telescopes. Mars becomes fainter than magnitude −2.0.

October 11, 2003 Mars recedes to 0.5 AU (74,800,000 km/ 46,480,000 miles) from Earth.

October 26, 2003 Mars' apparent disk diameter falls to 15 arcseconds.

November 3, 2003 Mars fades to magnitude −1.0.

December 13, 2003 Mars' apparent disk diameter is less than 10 arcseconds; detail is now best seen with large amateur telescopes. Mars is in Pisces.

December 19, 2003 Mars is at a distance of 1.0 AU (149,600,000 km/92,960,000 miles) from Earth.

December 21, 2003 Mars fades below magnitude 0.0.

December 29, 2003 Mars reaches quadrature, 90 degrees east of the Sun; it appears gibbous.

February 13, 2004 Mars is at a distance of 1.5 AU (224,400,000 km/139,440,000 miles) from Earth. Mars is now in Aries, moving steadily eastward and setting around midnight.

February 20, 2004 Mars' apparent disk diameter falls below 5 arcseconds.

February 21, 2004 Mars fades below magnitude +1.0.

March 26, 2004 The waxing crescent Moon is very close to Mars in the evening sky for European observers; in Northern Canada, the Moon occults Mars. Mars is in Taurus between the Hyades and Pleiades.

April 10, 2004 Mars is at a distance of 2.0 AU (299,200,000 km/185,920,000 miles) from Earth.

September 16, 2004 Mars is at conjunction with the Sun.

Mars ephemeris for 2003

The apparent position of Mars relative to the fixed star back ground can be defined by its location on the grid of right ascension (RA) and declination (Dec.) used to define the location of celestial objects. RA is equivalent to longitude on the sky, increasing eastward from the point of the northern hemisphere spring equinox (where the Sun crosses the celestial equator moving from south to north). It is usual to describe RA in hours and arcminutes: one hour is equivalent to 15 degrees, with each degree containing 60 arcminutes. For much of the time, except when close to opposition, Mars has a gradually increasing RA; only when it is in retrograde motion does Mars' RA decrease. Declination is measured north (+) or south (−) of the celestial equator in degrees and minutes. When well north of the celestial equator, Mars is high in the sky, meaning that it is well presented for observers at northerly latitudes on Earth. However, at its closest, as in 2003 for example, Mars is well south of the celestial equator, such that observers in southerly locations on Earth get the best view, with Mars at its highest when the apparent disk is largest.

For each apparition, an ephemeris (plural ephemerides) can be calculated in advance, showing Mars' apparent motion against the background stars. The ephemerides presented here – also summarized on the map – give positions for Mars at 5-day intervals. Information is for 00h UT on the date shown.

MARS EPHEMERIS FOR 2003–4					
Date 2003	RA	Dec.	Magnitude	Apparent Diameter	Distance (AU)
Mar 3	17h 54.8m	−23° 29.0m	+0.93	6″	1.51
Mar 8	18h 08.7m	−23° 33.7m	+0.86	6″	1.46
Mar 13	18h 22.5m	−23° 34.1m	+0.79	6″	1.42
Mar 18	18h 36.3m	−23° 30.3m	+0.72	6″	1.37
Mar 23	18h 50.1m	−23° 22.5m	+0.65	7″	1.33
Mar 28	19h 03.7m	−23° 10.6m	+0.58	7″	1.28
Apr 2	19h 17.3m	−22° 55.0m	+0.50	7″	1.24
Apr 7	19h 30.3m	−23° 35.9m	+0.42	7″	1.19
Apr 12	19h 43.9m	−22° 13.4m	+0.33	8″	1.15
Apr 17	19h 56.9m	−21° 48.1m	+0.25	8″	1.11
Apr 22	20h 09.7m	−21° 19.7m	+0.16	8″	1.07
Apr 27	20h 22.4m	−20° 48.9m	+0.06	9″	1.02
May 2	20h 34.8m	−20° 15.9m	−0.03	9″	0.98
May 7	20h 47.0m	−19° 41.1m	−0.13	9″	0.94
May 12	20h 58.8m	−19° 05.1m	−0.23	10″	0.90
May 17	21h 10.4m	−18° 28.0m	−0.34	10″	0.87
May 22	21h 21.6m	−17° 50.5m	−0.44	11″	0.83
May 27	21h 32.5m	−17° 12.9m	−0.55	11″	0.79
Jun 1	21h 42.9m	−16° 35.8m	−0.67	12″	0.75
Jun 6	21h 53.0m	−15° 59.9m	−0.79	13″	0.72
Jun 11	22h 02.5m	−15° 25.6m	−0.91	13″	0.69
Jun 16	22h 11.5m	−14° 53.5m	−1.04	14″	0.65
Jun 21	22h 19.9m	−14° 24.1m	−1.17	15″	0.62
Jun 26	22h 27.7m	−13° 58.3m	−1.30	15″	0.59
Jul 1	22h 34.7m	−13° 36.7m	−1.44	16″	0.56
Jul 6	22h 40.8m	−13° 20.0m	−1.58	17″	0.53
Jul 11	22h 46.1m	−13° 08.6m	−1.72	18″	0.51
Jul 16	22h 50.3m	−13° 03.0m	−1.87	19″	0.48
Jul 21	22h 53.4m	−13° 03.6m	−2.01	20″	0.46
Jul 26	22h 55.4m	−13° 10.7m	−2.16	21″	0.44
Jul 31	22h 56.0m	−13° 24.2m	−2.30	22″	0.42
Aug 5	22h 55.3m	−13° 43.6m	−2.44	23″	0.41
Aug 10	22h 53.4m	−14° 17.5m	−2.57	23″	0.39

MARS EPHEMERIS FOR 2003–4					
Date 2003	RA	Dec	Magnitude	Apparent Diameter	Distance (AU)
Aug 15	22h 50.3m	−14° 34.6m	−2.69	24″	0.38
Aug 20	22h 46.3m	−15° 03.0m	−2.79	24″	0.38
Aug 25	22h 41.4m	−15° 30.6m	−2.86	25″	0.37
Aug 30	22h 36.2m	−15° 55.0m	−2.88	25″	0.37
Sep 4	22h 30.9m	−16° 3.9m	−2.82	24″	0.38
Sep 9	22h 26.1m	−16° 25.6m	−2.72	24″	0.38
Sep 14	22h 21.9m	−16° 29.2m	−2.60	23″	0.39
Sep 19	22h 18.3m	−16° 24.7m	−2.47	22″	0.41
Sep 24	22h 16.6m	−16° 1.8m	−2.33	22″	0.42
Sep 29	22h 15.8m	−15° 51.0m	−2.18	21″	0.44
Oct 4	22h 16.3m	−15° 22.7m	−2.02	20″	0.47
Oct 9	22h 18.0m	−14° 47.8m	−1.87	19″	0.49
Oct 14	22h 20.9m	−14° 07.1m	−1.71	18″	0.52
Oct 19	22h 24.9m	−13° 21.1m	−1.56	17″	0.55
Oct 24	22h 29.8m	−13° 20.5m	−1.42	16″	0.57
Oct 29	22h 35.7m	−11° 35.6m	−1.27	15″	0.61
Nov 3	22h 42.3m	−10° 36.5m	−1.13	14″	0.64
Nov 8	22h 49.6m	−09° 34.2m	−0.99	13″	0.67
Nov 13	22h 57.4m	−08° 29.2m	−0.86	13″	0.71
Nov 18	23h 05.8m	−07° 21.5m	−0.73	12″	0.75
Nov 23	23h 14.6m	−06° 11.6m	−0.61	11″	0.79
Nov 28	23h 23.8m	−04° 59.5m	−0.49	11″	0.82
Dec 3	23h 33.3m	−03° 45.8m	−0.37	10″	0.86
Dec 8	23h 43.1m	−02° 30.5m	−0.26	10″	0.91
Dec 13	23h 53.2m	−01° 14.2m	−0.15	9″	0.95
Dec 18	00h 03.5m	+00° 02.9m	−0.05	9″	0.99
Dec 23	00h 14.1m	+01° 20.6m	+0.05	9″	1.03
Dec 28	00h 24.8m	+02° 36.7m	+0.15	8″	1.08
2004					
Jan 2	00h 35.8m	+03° 56.8m	+0.24	8″	1.12
Jan 7	00h 46.9m	+05° 14.7m	+0.33	8″	1.17
Jan 12	00h 58.2m	+06° 32.0m	+0.42	7″	1.21
Jan 17	01h 09.6m	+07° 46.5m	+0.50	7″	1.25
Jan 22	01h 21.2m	+09° 04.0m	+0.58	7″	1.30
Jan 27	01h 32.9m	+10° 18.3m	+0.66	6″	1.35
Feb 1	01h 44.8m	+11° 30.9m	+0.73	6″	1.39
Feb 6	01h 56.9m	+12° 41.8m	+0.80	6″	1.44
Feb 11	02h 09.1m	+13° 50.5m	+0.87	6″	1.49
Feb 16	02h 21.5m	+14° 56.9m	+0.87	6″	1.53

MARS IN 2005

The very favorable apparition that brings Mars to a perihelic opposition in August 2003 takes just over a year to wind down. Mars finally arrives at conjunction, on the far side of the Sun, on September 16, 2004. Thereafter it remains dim and distant for a long interval. It gradually creeps into the eastern predawn sky during the closing weeks of 2004, but remains far from obvious, at magnitude +1.6, at the year's end. By late March 2005, Mars is a bit more conspicuous, reaching magnitude +1.0, and in mid-April the distance between Earth and the Red Planet begins to shrink.

Many observers will start to pick up Mars for the 2005 apparition during late June 2005, as it brightens to magnitude 0 while among the stars of Pisces, rising in the small hours. Quadrature, 90 degrees west of the Sun, is reached against the background of Aries on July 13. Now less than one astronomical unit (AU) away, Mars reveals a disk of 10 arcseconds' diameter, which is sufficiently large for a medium-aperture telescope to show some detail. Mars' gibbous phase should be obvious, with the evening terminator on the preceding side of the disk. Those using telescopes with apertures of 200 mm or greater might be able to discern white clouds forming in the Martian atmosphere over those parts of the planet close to, or just beyond, the day–night line.

By September, Mars is rising in late evening, becoming a prominent object, at magnitude −1.0: Mars is the brightest object in the midnight sky at this time. Mid-month sees the apparent disk diameter in excess of 15 arcseconds, allowing surface markings to be seen even in small telescopes.

Mars' direct, eastward motion slows and comes to a halt in early October, about 10 degrees west of the Pleiades star cluster in the sky. Fitting comfortably into the same field of view in a standard 50 mm SLR camera lens, the combination of Mars and the Pleiades makes an attractive composition for photography; observers with 135 mm telephoto lenses should also be able to fit Mars and the Pleiades into the same shot, for more of a "close-up" view.

As October progresses, Earth closes on Mars, and the Red Planet brightens to magnitude −2, reaching its maximum apparent diameter of 20 arcseconds between October 23 and November 7. Closest approach is between October 28 and 31, a week or so ahead of opposition; this situation is quite common, and results from the inclination of Mars' orbit relative to that of Earth. At its closest, Mars is 0.464 AU (69.4 million kilometers/43.1 million miles) away, which is about one and a quarter times its distance at opposition in 2003.

Opposition, in eastern Aries, occurs on November 7, when Mars has an apparent magnitude of −2.33. With the Moon a thin waxing crescent setting in early evening, and Venus and Jupiter also setting before the night is fully dark, Mars dominates the midnight skies of early November.

Moving into the evening sky, Mars remains prominent through the rest of November, fading only slowly. In early December, the apparent disk shrinks to 15 arcseconds' diameter. The waxing gibbous Moon (two days from Full) is nearby on the night of December 11–12. Mars' retrograde motion stops on this date, and the planet resumes its eastward progress once more.

By December 19, Mars has faded below magnitude −1, and over the coming weeks the planet's brightness drops quite rapidly. Come late January 2006, Mars is fainter than magnitude 0, shows a disk smaller than 10 arcseconds across, and is more than one astronomical unit away. For observers using smaller telescopes, the show is as good as over!

Quadrature, 90 degrees east of the Sun, is reached on February 18, when Mars appears close in the sky to the Pleiades. Observers using large telescopes will see the dawn terminator on the following side of Mars' 7-arcsecond disk, and may be able to detect early-morning hazes.

In April, Mars fades still further and becomes fairly inconspicuous in the early evening western sky. However, it has a couple of interesting close approaches – known as appulses – in our line of sight to objects in the vastly more remote stellar background. Around April 17–18, it passes less than one degree north of the star cluster M35. On May 2, Mars is 0.4 degrees (24 arcminutes; four-fifths the Moon's apparent diameter) from the star Epsilon Geminorum.

On April 7–8, 1976, Mars actually passed in front of Epsilon Geminorum, a magnitude +3.0 star, occulting it. The event came some months after an aphelic opposition, with the planet at magnitude +1.1 and showing a disk of 6 arcseconds' diameter. Mars' approach to the star was followed by many amateur observers using large telescopes. As it was occulted, Epsilon Geminorum flickered and faded due to effects of the Martian atmosphere. The occultation lasted 8 minutes. At mid-occultation, professional photometric observations suggested a brief flash of brightening, which was perhaps caused by refraction of the star's light in Mars' atmosphere.

From mid-April, the apparition continues its usual slow wind down, with Mars becoming ever more deeply lost into the evening twilight. A passage in front of the Praesepe open star cluster (M44) in Cancer in mid-June 2006 is essentially unobservable in the bright sky, as is a conjunction with Saturn a few day later, on June 17. By July, Mars has moved into Leo and is unlikely to be visible at all, arriving finally at conjunction on the far side of the Sun on October 23, 2006.

Highlights for telescopic observers in 2005

At the 2005 opposition, Mars will be much more favorably placed in the sky for observers at northerly latitudes on Earth than it was in 2003. From southern England, Mars will culminate a healthy 55 degrees above the horizon, 50 degrees in Scotland. Observers at the latitude of New York (40°N) will see Mars at its highest 65 degrees above the horizon. These higher-elevation views are less likely to be prone to the effects of poor seeing than those at the previous opposition, when observers in the northern hemisphere saw Mars low down through a thick wedge of unsteady air close to the horizon.

At opposition in November 2005, it will be late summer in Mars' southern hemisphere. The south polar cap – again well presented toward Earth – will be at its minimum extent. The separation and fragmentation of the Novus Mons ("Mountains of Mitchell") will occur in early July, well ahead of opposition. At this time the disk diameter is only 10 arcseconds, so large telescopes under ideal conditions will be required to reveal much detail.

As late summer advances into autumn in Mars' southern hemisphere, the south polar cap should become shrouded in haze, appearing featureless on most nights. Any dust storm activity is likely to be subsiding by the time of opposition, in early November. Yellow clouds may be seen quite frequently early in the apparition, during July, August and September.

For observers in western Europe, the dark triangle of Syrtis Major will be presented on the center of Mars' visible hemisphere during the evening hours in late October and the first week of December. Observers in the eastern United States will see Syrtis Major well on evenings during early November and the middle part of December. West-coast United States locations get their best evening views five to seven days later. In Australia and New Zealand, prime evening viewing times for the Syrtis Major come in mid-October and the third week of November.

The "Eye of Mars," Solis Planum, is well placed for evening observation from western Europe in the second week of October, the third week of November and the closing week of December. Observers in the eastern United States will see this part of Mars in their evening views during the third week of October and last week

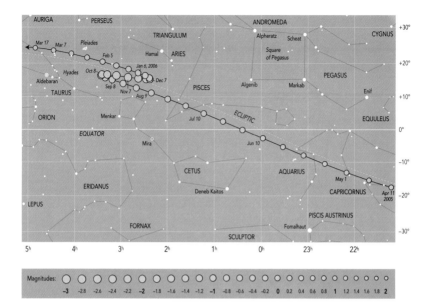

▲ *The apparent path of Mars against the star background in 2005–6. Larger circles indicate greater brightness, as in the magnitude scale at bottom.*

of November. On the western side of North America, Solis Planum is in view on evenings during the last week of October and in early December. Observers in New Zealand and Australia will see Solis Planum well placed in their evening views of Mars during early November, and again in the second week of December.

Mars 2005 apparition timetable

This timetable describes key moments in the apparition. To give a full picture of the planet's movement, it begins and ends with Mars at conjunction.

September 16, 2004 Mars is at conjunction with the Sun.

October 2004 to January 2005 Mars emerges slowly into the morning sky, passing through Libra and Scorpius. By December 2004, Mars is rising 2.5 hours ahead of the Sun, but is not particularly prominent, at magnitude +1.6.

February 5, 2005 Mars' distance falls below 2.0 AU (299,200,000 km/185,920,000 miles) from Earth.

February 23, 2005 Mars' apparent disk diameter increases to 5 arcseconds.

March 22, 2005 Mars brightens to magnitude +1.0.

April 15, 2005 Mars' distance falls below 1.5 AU (224,400,000 km/139,440,000 miles) from Earth.

June 27, 2005 Mars, on the Pisces/Cetus border, is now brighter than magnitude 0.0, rising in early morning hours.

July 1, 2005 Mars is now 1.0 AU (149,600,000 km/92,960,000 miles) from Earth.

July 13, 2005 Mars' apparent disk diameter reaches 10 arcseconds; larger amateur telescopes should by now show some detail. Mars is at quadrature, 90 degrees west of the Sun, and shows a gibbous phase. It rises close to midnight, among the stars of Aries.

September 1, 2005 Mars is brighter than magnitude −1.0, rising in late evening.

September 9, 2005 Mars' apparent disk diameter reaches 15 arcseconds; even modest amateur telescopes now reveal some detail.

October 2, 2005 Direct (eastward) motion ceases; Mars starts to move retrograde, about 10 degrees west of the Pleiades.

October 9, 2005 Mars' distance falls below 0.5 AU (74,800,000 km/46,480,000 miles).

October 15, 2005 Mars brightens to magnitude −2.0.

October 23, 2005 Mars' apparent disk diameter reaches 20 arcseconds; between now and November 7 it is at its maximum apparent size for the apparition.

October 28–31, 2005 Mars is at its closest for the apparition, at a distance of 0.464 AU (69,400,000km/43,100,000 miles).

November 7, 2005 Mars is at opposition in Aries; it is at its brightest for the apparition, at magnitude −2.33, and is visible throughout the hours of darkness. Its disk diameter falls below 20 arcseconds.

November 19, 2005 Mars is fading, and is now magnitude −2.0. Its distance from Earth increases to over 0.5 AU (74,800,000 km/ 46,480,000 miles).

December 7, 2005 Mars' apparent diameter falls below 15 arc-seconds.

December 11–12, 2005 The waxing gibbous Moon is close to Mars in the late evening for western European observers. Retrograde motion ends; Mars begins direct (eastward) motion relative to the star background once more.

December 19, 2005 Mars is fainter than magnitude −1.0.

January 19, 2006 Mars' apparent disk diameter is less than 10 arc-seconds.

January 23, 2006 Mars is now fainter than magnitude 0.0.

January 26, 2006 Mars is at a distance of 1.0 AU (149,600,000 km/ 92,960,000 miles) from Earth.

February 18, 2006 Mars is at quadrature, 90 degrees east of the Sun, and shows a gibbous phase. It is close to the Pleiades, in Taurus.

March 17, 2006 Mars is at a distance 1.5 AU (224,400,000 km/ 139,440,000 miles) from Earth. It fades below magnitude +1.5.

March 24, 2006 Mars' disk diameter falls to 5 arcseconds. Only large telescopes will show much detail now.

April 17–18, 2006 Mars passes less than one degree to the north of the star cluster M35 in Gemini.

May 2, 2006 Mars, at magnitude +1.5, is 24 arcminutes south of the magnitude +3.0 star Epsilon Geminorum.

June 15–17, 2006 Mars, magnitude +1.8, passes across the Praesepe star cluster (M44) in Cancer, low in the evening sky.

June 17, 2006 Mars is 35 arcminutes north of magnitude +0.4 Saturn.

October 23, 2006 Mars is at conjunction with the Sun.

For each apparition, an ephemeris (plural ephemerides) can be calculated in advance, showing Mars' apparent motion against the background stars. The ephemerides presented here – also summarized on the map – give positions for Mars at 5-day intervals, as well as information on its apparent diameter and distance from Earth. Information is for 00h UT on the date shown.

MARS EPHEMERIS FOR 2005–6					
Date 2005	RA	Dec	Magnitude	Apparent Diameter	Distance (AU)
Apr 11	21h 12.9m	−17° 26.5m	+0.83	6″	1.53
Apr 16	21h 27.4m	−16° 25.0m	+0.78	6″	1.49
Apr 21	21h 41.8m	−15° 20.2m	+0.73	6″	1.46
Apr 26	21h 56.1m	−14° 12.2m	+0.68	6″	1.42
May 1	22h 10.2m	−13° 01.5m	+0.63	6″	1.39
May 6	22h 24.2m	−11° 48.4m	+0.58	6″	1.35
May 11	22h 38.0m	−10° 33.2m	+0.53	7″	1.32
May 16	22h 51.7m	−09° 16.5m	+0.47	7″	1.29
May 21	23h 05.8m	−07° 58.5m	+0.42	7″	1.26
May 26	23h 18.7m	−06° 39.4m	+0.37	7″	1.22
May 31	23h 31.9m	−05° 19.8m	+0.31	7″	1.19
Jun 5	23h 45.1m	−03° 59.9m	+0.26	8″	1.16
Jun 10	23h 58.1m	−02° 40.3m	+0.20	8″	1.13
Jun 15	00h 11.0m	−01° 21.1m	+0.14	8″	1.10
Jun 20	00h 23.7m	−00° 02.9m	+0.08	8″	1.07
Jun 25	00h 36.3m	+01° 14.1m	+0.02	9″	1.04
Jun 30	00h 48.8m	+02° 29.7m	−0.04	9″	1.01
Jul 5	01h 01.1m	+03° 43.4m	−0.10	9″	0.98
Jul 10	01h 13.2m	+04° 40.3m	−0.17	9″	0.95
Jul 15	01h 25.1m	+06° 03.8m	−0.23	10″	0.92
Jul 20	01h 36.8m	+07° 10.1m	−0.30	10″	0.90
Jul 25	01h 48.3m	+08° 13.3m	−0.37	10″	0.87
Jul 30	01h 59.5m	+09° 13.5m	−0.44	11″	0.84
Aug 4	02h 10.3m	+10° 10.2m	−0.52	11″	0.81
Aug 9	02h 20.7m	+11° 03.2m	−0.60	11″	0.79
Aug 14	02h 30.7m	+11° 52.5m	−0.68	12″	0.76
Aug 19	02h 40.1m	+12° 38.0m	−0.77	12″	0.73
Aug 24	02h 50.3m	+13° 19.8m	−0.86	13″	0.70
Aug 29	02h 57.2m	+13° 57.5m	−0.95	13″	0.68
Sep 3	03h 04.6m	+14° 31.3m	−1.05	14″	0.65
Sep 8	03h 11.1m	+15° 01.0m	−1.16	14″	0.63
Sep 13	03h 16.6m	+15° 26.8m	−1.26	15″	0.60

MARS EPHEMERIS FOR 2005–6					
Date 2005	RA	Dec	Magnitude	Apparent Diameter	Distance (AU)
Sep 18	03h 20.9m	+15° 48.8m	−1.37	16″	0.58
Sep 23	03h 24.1m	+16° 06.8m	−1.49	16″	0.56
Sep 28	03h 25.8m	+16° 20.9m	−1.60	17″	0.54
Oct 3	03h 26.2m	+16° 30.7m	−1.72	18″	0.52
Oct 8	03h 24.9m	+16° 36.9m	−1.84	18″	0.50
Oct 13	03h 22.1m	+16° 38.8m	−1.95	19″	0.49
Oct 18	03h 17.9m	+16° 36.8m	−2.05	19″	0.48
Oct 23	03h 12.4m	+16° 30.8m	−2.15	20″	0.47
Oct 28	03h 05.9m	+16° 21.3m	−2.23	20″	0.46
Nov 2	02h 58.7m	+16° 08.9m	−2.29	20″	0.46
Nov 7	02h 51.2m	+15° 54.8m	−2.33	19″	0.47
Nov 12	02h 44.0m	+15° 40.5m	−2.22	19″	0.48
Nov 17	02h 37.5m	+15° 27.6m	−2.06	19″	0.49
Nov 22	02h 31.8m	+15° 17.2m	−1.93	18″	0.51
Nov 27	02h 27.3m	+15° 10.5m	−1.76	17″	0.53
Dec 2	02h 24.2m	+15° 08.3m	−1.59	16″	0.56
Dec 7	02h 22.4m	+15° 11.2m	−1.42	15″	0.59
Dec 12	02h 22.0m	+15° 19.3m	−1.25	15″	0.62
Dec 17	02h 22.9m	+15° 32.3m	−1.09	14″	0.66
Dec 22	02h 25.1m	+15° 49.9m	−0.92	13″	0.69
Dec 27	02h 28.3m	+16° 11.7m	−0.76	12″	0.73
2006					
Jan 1	02h 32.6m	+16° 37.1m	−0.61	12″	0.78
Jan 6	02h 37.8m	+17° 05.5m	−0.46	11″	0.82
Jan 11	02h 43.8m	+17° 36.4m	−0.32	10″	0.86
Jan 16	02h 50.6m	+18° 09.2m	−0.19	10″	0.91
Jan 21	02h 58.1m	+18° 43.2m	−0.06	9″	0.96
Jan 26	03h 06.2m	+19° 17.9m	+0.06	9″	1.00
Jan 31	03h 14.8m	+19° 53.0m	+0.18	8″	1.05
Feb 5	03h 23.9m	+20° 27.9m	+0.29	8″	1.10
Feb 10	03h 33.5m	+21° 02.4m	+0.39	8″	1.15
Feb 15	03h 43.5m	+21° 35.6m	+0.49	7″	1.20
Feb 20	03h 53.9m	+22° 07.5m	+0.59	7″	1.25
Feb 25	04h 04.7m	+22° 37.6m	+0.68	7″	1.30
Mar 2	04h 15.7m	+23° 05.7m	+0.77	6″	1.35
Mar 7	04h 27.1m	+23° 31.5m	+0.85	6″	1.40
Mar 12	04h 38.7m	+24° 14.7m	+0.92	6″	1.45
Mar 17	04h 50.5m	+24° 14.7m	+1.00	6″	1.50
Mar 22	05h 02.5m	+24° 31.8m	+1.06	6″	1.55

MARS IN 2007–8

Following conjunction in October 2006, Mars creeps back into the early morning sky, staying well south of the celestial equator as it moves against the background stars of Sagittarius, Capricornus and Aquarius. Not until September 2007 does Mars pull out to quadrature, 90 degrees west of the Sun, just north of Zeta Tauri, the Bull's southerly horn near the Taurus/Gemini border. It is probably around this time that most casual observers will begin to notice the Red Planet as a rather more prominent object in the early morning sky, shining at magnitude 0. The planet's phase at this time is gibbous, with the evening terminator well on to the disk at the preceding limb in the view from Earth. Earth and Mars are one astronomical unit apart on September 25, 2007.

In early October, Mars is close to the open star cluster M35 in Gemini. This pairing will be attractive in binoculars, and can be captured photographically in driven exposures with a 135 mm or similar telephoto lens. From the second week of October, the apparent disk diameter exceeds 10 arcseconds, and larger amateur telescopes should provide rewarding views of Mars' albedo features in the early morning hours.

This is a good apparition for observers at higher northerly latitudes, with Mars coming to opposition near the "top" of the ecliptic. Throughout the interval from October to February, when at its best, Mars culminates at an altitude of about 65 degrees above the southern horizon for observers in the United Kingdom. At the latitude of New York, Mars reaches a maximum altitude close to 75 degrees. Observers in western Europe and North America should enjoy reasonable views: despite having a relatively small apparent diameter for much of the apparition, Mars is well above the worst of any atmospheric turbulence.

By early November, Mars has brightened to magnitude −1, becoming the brightest object in the late evening sky (outshone later in the night only by Sirius). On November 16, direct (eastward) motion relative to the star background slows and comes to a halt a few degrees south of Epsilon Geminorum, as Earth continues to catch up to Mars. Hereafter, Mars is moving retrograde from Gemini toward Taurus.

At this apparition, Mars reaches a maximum apparent disk diameter of 15 arcseconds, which is 60% of its size at the perihelic opposition of 2003. This is sufficient to allow good views in, say, a 114 mm refractor or 150 mm reflector. Mars' disk stays close to this apparent size through December and into the first week of January 2008.

Mars is at its closest on December 18 and 19, 2007, slightly ahead of opposition, which occurs on Christmas Eve in western Gemini near the constellation's border with Taurus. Mars, at this time, is about as far north as it can be on the celestial sphere. The planet reaches its peak brightness for the apparition on December 23 to 24, at magnitude −1.64. On this occasion, though brighter than any of the stars, Mars at its best has to contend with a bright Full Moon close by in the midnight sky. Nonetheless, the planet's strong red color makes it a standout object at this time.

After January 7, 2008, Mars' apparent disk diameter begins to drop quite rapidly. Observers using small telescopes should make the most of any viewing opportunities during January, since Mars will have shrunk to a 10-arcsecond disk by mid-February. Direct, eastward motion resumes on January 31, with Mars close to Taurus' northern horn.

As it recedes – or, more precisely, Earth speeds by on its faster, inner orbit – Mars fades quickly. By late January, Mars is fainter than magnitude −1.0, and by the third week of February it has faded below magnitude 0.

From here, the apparition winds down steadily, and only those with larger telescopes will see much detail from late February 2008 onward. Quadrature, 90 degrees east of the Sun, occurs on March 30, when the morning terminator is well on to the following part of the disk, giving Mars a distinct gibbous phase. At this time, Mars passes close to the third-magnitude star Epsilon Geminorum.

In mid-April, Mars' brightness falls to magnitude +1.0, and its disk diameter to only 5 arcseconds; this month marks the end of the apparition for most telescope users.

The Red Planet has its usual lingering end to the naked eye apparition, as a dim red spark in the evening twilight. On May 23 and 24, Mars passes in front of the Praesepe open star cluster in Cancer, a spectacle which will be rather muted in the bright sky. From July onward, Mars becomes very much lost in the twilight, and finally reaches conjunction beyond the Sun on December 5, 2008.

Highlights for telescopic observers in 2007–8

At this apparition, Mars is again presented with it southern hemisphere tilted toward Earth. Around the time of opposition, it will be mid-autumn in the Martian southern hemisphere, and observers can expect to see the south polar cap shrouded in haze. Any substantial dust storm/yellow cloud activity should have died down by the time Mars comes into favorable view during October. White clouds are, however, quite likely to be seen; observers with large telescopes should keep an eye on the terminator, especially in the weeks around quadrature, for dawn and evening hazes extending over the day–night line.

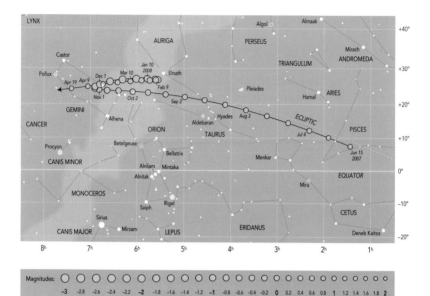

▲ *The apparent path of Mars against the star background in 2007–8. Larger* circles indicate greater brightness, as in the magnitude scale at bottom.

With yellow cloud activity probably at a relatively low level, observers with medium-sized telescopes might hope for reasonably clear views of the albedo features, albeit on a smaller-than-ideal disk.

From the British Isles, observers can expect to see the dark Syrtis Major near the central meridian on evenings in late November, and in the interval around New Year 2008. Observers in eastern North America have ideal evening presentation of Syrtis Major in the first week of December and second week of January, while those in the western United States should see it best in the second weeks of December and January. From New Zealand and Australia, Syrtis Major is on Mars' visible hemisphere during evenings in the second week of November and third week of December, but the planet's low elevation from southerly terrestrial latitudes may pose some problems for clear viewing.

Solis Planum and its environs are on the Earth-facing hemisphere in evenings during the second weeks of November and December for UK-based observers. The eastern US has good views in mid-November and around Christmas 2007, while western North American observers see this part of Mars best in the third week of November and the last week of December. In Australia and New Zealand, prime viewing for Solis Planum can be had in early December and the second week of January.

It may be worth looking for orographic cloud forming over the Tharsis ridge, on the preceding side of Mars' visible disk, a week or so after these ideal "windows" for viewing Solis Planum.

2007–8 apparition timetable

This timetable describes key moments in the apparition. To give a full picture, it begins and ends with Mars at conjunction.

October 23, 2006 Mars is at conjunction, on the far side of the Sun.

November 2006 to April 2007 Mars emerges only slowly into the morning sky, passing through Sagittarius, Capricornus and Aquarius.

April 29, 2007 Mars is 0.7 degrees from Uranus, poorly placed in the morning sky. Mars is at magnitude +1.0 and has a disk diameter of 5 arcseconds.

June 27, 2007 Mars, in Aries, is now less than 1.5 AU (224,400,000 km/139,440,000 miles) from Earth.

September 18, 2007 Mars reaches quadrature, 90 degrees west of the Sun, in Taurus.

September 25, 2007 Mars closes to within 1 AU (149,600,000 km/92,960,000 miles) of Earth. It brightens to magnitude 0.0.

October 4, 2007 Mars passes about one degree south of the open star cluster M35 in Gemini.

October 6, 2007 Mars' apparent disk diameter reaches 10 arcseconds.

November 16, 2007 Direct motion stops; Mars begins to move retrograde (westward), a few degrees south of Epsilon Geminorum.

November 17, 2007 Mars brightens to magnitude −1.0.

November 30, 2007 Mars' apparent disk diameter reaches 15 arcseconds. This is as large as Mars will become this apparition, but it remains so until early January.

December 18–19, 2007 Mars is at its closest for the apparition, at a distance of 0.594 AU (88,000,000 km/55,000,000 miles) from Earth.

December 23–24, 2007 Mars is at its brightest this apparition, at magnitude −1.64.

December 24, 2007 Mars is at opposition in western Gemini. The Full Moon is nearby.

January 7, 2008 Mars' disk diameter falls below 15 arcseconds.

January 18, 2008 Mars is fainter than magnitude −1.0.

January 19–20, 2008 The waxing gibbous Moon is fairly close to the north of Mars.

January 31, 2008 Mars, now close in the sky to Beta Tauri, resumes direct (eastward) motion relative to the star background.

February 22, 2008 Mars' magnitude falls to 0.0.

February 24, 2008 Mars' apparent disk diameter falls below 10 arcseconds.

February 28, 2008 Mars is at a distance of more than 1 AU (149,600,000 km/92,960,000 miles) from Earth.

March 30, 2008 Mars reaches quadrature, 90 degrees east of the Sun. Mars (magnitude +0.8) is 16 arcminutes from Epsilon Geminorum (magnitude +3.0).

April 14, 2008 Mars fades to magnitude +1.0.

April 25, 2008 Mars' apparent diameter decreases to 5 arcseconds.

May 23–24, 2008 Mars passes in front of the Praesepe open star cluster (M44) in Cancer.

June 15, 2008 Mars is now a very distant 2 AU (299,200,000 km/185,920,000 miles) from Earth, in Leo.

July–September 2008 Mars becomes lost in evening twilight among the stars of Virgo.

December 5, 2008 Mars is at conjunction, on the far side of the Sun.

The ephemerides presented here – also summarized on the map – give positions for Mars at 5-day intervals. Information is for 00h UT on the date shown.

MARS EPHEMERIS FOR 2007–8					
Date 2007	RA	Dec	Magnitude	Apparent Diameter	Distance (AU)
Jun 15	01h 26.4m	+07° 26.6m	+0.78	6″	1.56
Jun 20	01h 40.2m	+08° 40.7m	+0.76	6″	1.53
Jun 25	01h 53.9m	+10° 06.3m	+0.73	6″	1.51
Jun 30	02h 07.7m	+11° 22.0m	+0.71	6″	1.48
Jul 4	02h 18.7m	+12° 20.3m	+0.69	6″	1.46
Jul 9	02h 32.4m	+13° 20.3m	+0.66	6″	1.44
Jul 14	02h 46.2m	+14° 36.8m	+0.63	6″	1.42
Jul 19	02h 59.8m	+15° 39.5m	+0.60	6″	1.39
Jul 24	03h 13.6m	+16° 38.2m	+0.57	6″	1.36
Jul 29	03h 27.2m	+17° 32.6m	+0.54	6″	1.34
Aug 3	03h 40.7m	+18° 23.5m	+0.51	6″	1.31
Aug 8	03h 54.2m	+19° 09.8m	+0.47	7″	1.29
Aug 13	04h 07.5m	+19° 51.9m	+0.44	7″	1.26
Aug 18	04h 20.7m	+20° 29.7m	+0.40	7″	1.20
Aug 23	04h 33.5m	+21° 03.3m	+0.36	7″	1.17
Aug 28	04h 46.4m	+21° 32.8m	+0.31	7″	1.14
Sep 2	04h 58.9m	+21° 58.3m	+0.27	8″	1.14
Sep 7	05h 11.1m	+22° 20.2m	+0.21	8″	1.12
Sep 12	05h 23.0m	+22° 38.3m	+0.16	8″	1.08
Sep 17	05h 34.4m	+22° 53.6m	+0.10	8″	1.05
Sep 22	05h 45.4m	+23° 05.9m	+0.04	9″	1.02
Sep 27	05h 55.8m	+23° 15.7m	−0.02	9″	0.99
Oct 2	06h 05.8m	+23° 23.8m	−0.09	9″	0.96
Oct 7	06h 15.1m	+23° 29.7m	−0.17	10″	0.93
Oct 12	06h 23.5m	+23° 34.8m	−0.25	10″	0.90
Oct 17	06h 31.3m	+23° 39.5m	−0.33	10″	0.86
Oct 22	06h 38.1m	+23° 44.4m	−0.42	11″	0.83
Oct 27	06h 43.8m	+23° 49.9m	−0.52	11″	0.80
Nov 1	06h 48.6m	+23° 56.8m	−0.62	12″	0.77
Nov 6	06h 52.0m	+24° 05.3m	−0.73	12″	0.74
Nov 11	06h 54.2m	+24° 16.2m	−0.84	13″	0.71
Nov 16	06h 54.8m	+24° 29.5m	−0.95	13″	0.69
Nov 21	06h 53.9m	+24° 45.3m	−1.07	14″	0.66
Nov 26	06h 51.5m	+25° 03.3m	−1.18	14″	0.64
Dec 1	06h 47.5m	+25° 23.0m	−1.29	15″	0.62

MARS EPHEMERIS FOR 2007–8						
Date 2007	RA	Dec	Magnitude	Apparent Diameter	Distance (AU)	
Dec 6	06h 41.8m	+25° 43.5m	−1.39	15″	0.61	
Dec 11	06h 34.9m	+26° 03.5m	−1.48	15″	0.60	
Dec 16	06h 27.0m	+26° 21.8m	−1.56	15″	0.59	
Dec 21	06h 18.5m	+26° 36.9m	−1.62	15″	0.59	
Dec 26	06h 09.6m	+26° 48.3m	−1.62	15″	0.59	
Dec 31	06h 01.1m	+26° 55.4m	−1.52	15″	0.61	
2008						
Jan 5	05h 53.2m	+26° 58.6m	−1.40	15″	0.62	
Jan 10	05h 46.4m	+26° 58.5m	−1.26	14″	0.64	
Jan 15	05h 40.9m	+26° 55.8m	−1.11	14″	0.67	
Jan 20	05h 36.9m	+26° 51.7m	−0.96	13″	0.69	
Jan 25	05h 34.5m	+26° 46.9m	−0.81	12″	0.73	
Jan 30	05h 33.5m	+26° 42.1m	−0.65	12″	0.76	
Feb 4	05h 33.9m	+26° 37.3m	−0.51	11″	0.80	
Feb 9	05h 35.8m	+26° 33.0m	−0.36	11″	0.84	
Feb 14	05h 38.8m	+26° 28.9m	−0.12	10″	0.89	
Feb 19	05h 42.9m	+26° 25.1m	−0.05	10″	0.93	
Feb 24	05h 48.1m	+26° 21.1m	+0.04	9″	0.98	
Feb 29	05h 54.0m	+26° 16.6m	+0.16	9″	1.02	
Mar 5	06h 00.8m	+26° 11.4m	+0.28	8″	1.07	
Mar 10	06h 08.3m	+26° 05.2m	+0.39	8″	1.12	
Mar 15	06h 16.3m	+25° 57.5m	+0.49	8″	1.17	
Mar 20	06h 24.9m	+25° 48.2m	+0.59	7″	1.22	
Mar 25	06h 34.1m	+25° 36.8m	+0.68	7″	1.27	
Mar 30	06h 43.5m	+25° 23.3m	+0.77	7″	1.32	
Apr 4	06h 53.3m	+25° 07.6m	+0.85	6″	1.37	
Apr 9	07h 03.4m	+24° 49.2m	+0.93	6″	1.42	
Apr 14	07h 13.8m	+24° 27.9m	+1.00	6″	1.47	
Apr 19	07h 24.6m	+24° 03.8m	+1.07	6″	1.51	
Apr 24	07h 35.2m	+23° 36.9m	+1.13	6″	1.56	

Following the 2007–8 apparition are a couple of returns where Mars is rather far away, and still more reluctant to give up its secrets. The oppositions of January 2010 and March 2013 see Mars attain a maximum apparent diameter of only 14 arcseconds, and on neither occasion will the planet be particularly favorable for observation with small telescopes. Not until July 2017 will there be another perihelic opposition to compare with that of 2003. Good viewing opportunities for Mars are sufficiently rare that observers are well advised to make the most of them!

EARLY EXPLORERS

Together with Mercury, Venus, Jupiter and Saturn, Mars is one of the five naked eye planets known since ancient times. The movements of these "wanderers" against the fixed star background were studied as far back as Babylonian times (2000 BC). In the cosmology of the classical Greeks, Aristotle's geocentric model, in which the planets and Sun orbited Earth in perfect circles, held sway. A problem for this world view, however, was the retrograde motion shown for several weeks of the year by each of the superior planets; Mars' retrograde motion, as we have seen, is especially pronounced.

In around 220 BC, Apollonius of Perga (262–190 BC) proposed a solution, in which he introduced smaller circular motions – known

▼ In the geocentric world view adopted by the classical Greeks, the planets and the Sun traveled around the Earth in perfectly circular orbits. The apparent retrograde motion shown by planets close to opposition was a problem for the system. Astronomers got round this problem by proposing that planets also moved in small circles – epicycles – superimposed on their orbits.

as epicycles – centered on the larger circle (the deferent) that marked each planet's orbit around the Earth. As it traveled around the epicycle, Mars would spend some time apparently moving backward along its orbit in this model.

Not all the classical Greeks adhered to the geocentric model. Aristarchus of Samos (310–230 BC), for example, put forward a heliocentric view, with Earth and the other planets orbiting the Sun.

However, bolstered in part by religious dogma, the geocentric Universe became the only acceptable model and remained so well into the 16th and 17th centuries.

Moving the Earth

As the centuries advanced, ever-increasing numbers of epicycles were required for the geocentric model of the Solar System to account for the observed motions of the planets. In the 16th century, the heliocentric theory, with the Sun replacing Earth at the center of things, was brought to the fore as an alternative, most notably by the publication in the year of his death of *De revolutionibus orbium coelestium* ("On the motions of the heavenly spheres") by the Polish astronomer Nicolas Copernicus (1473–1543). Copernicus' ideas were slow to gain acceptance, however, and the geocentric view continued to dominate.

Mars played an important part in the eventual establishment of the heliocentric view as the correct model for the Solar System. This role stemmed from the work of the Danish astronomer Tycho Brahe (1546–1601), surely the greatest of pre-telescopic observers. Using specially designed mural quadrants and other instruments at his observatory on the island of Hven in the Copenhagen Sound, Brahe made very precise measurements of the positions of the stars and planets, hoping to use them to prove his own modified Solar System view (now called the Tychonian model by historians of astronomy), in which the other planets orbited the Sun, which in turn orbited Earth.

After leaving Hven in 1597, Tycho Brahe moved to Prague, where several astronomers studied under him, among them a young Austrian, Johannes Kepler (1571–1630). Kepler was to inherit Brahe's high-precision observations, and he went on to use them to develop, finally, an acceptable heliocentric model for the Solar System. Instead of having circular orbits, planets move around the Sun in ellipses, thus accounting for their observed motions. Key to the development of Kepler's theories, summarized in his three laws of planetary motion, were Brahe's very accurate positional measurements of Mars against the fixed star background. Rather than proving his own alternative view of how the Solar System worked, Tycho Brahe left behind the precise data necessary to cement the heliocentric model.

Early telescopic views

In 1609, not long after Kepler's work on planetary motion, the Italian scientist Galileo Galilei (1564–1642) turned the recently invented telescope heavenward, making a string of discoveries that would forever change our view of the Solar System. For the first time, the other planets were revealed as worlds in their own right: Venus showed phases like those of the Moon, while Jupiter had its own collection of four bright satellites. Galileo's small, simple telescopes offered a magnification of ×20, which was enough to reveal the craters of the Moon. They suffered terribly, however, from chromatic aberration and other optical defects and, hardly surprisingly, showed nothing to inspire further interest when turned toward Mars. For some time, the Red Planet remained largely neglected by early telescopic observers.

Others may have glimpsed them earlier, but the credit for the first detection and description of Mars' dark markings is usually given to the Dutch astronomer Christiaan Huygens (1629–95) in 1659. By this time, the optical quality of telescope lenses was beginning to improve. One solution to the problem of chromatic aberration, employed by Huygens and others, was to build instruments with objective lenses of extremely long focal length. Some of these instruments grew into the vast, unwieldy "aerial" telescopes, in which the object glass, mounted on a tall pole, could be tens of meters from the eyepiece position.

In 1655, using a 51 mm aperture lens of 3.2 meters focal length and a magnification of ×50, Huygens discovered the true nature of Saturn's rings. Four years later, on November 28, 1659, Huygens examined Mars, which was then close to opposition in Taurus, east of the Hyades, high in the sky with an apparent diameter of 17 arcseconds. He found a V-shaped marking on the disk: his sketch marks the first description of what we now know as Syrtis Major.

▶ *The Dutch astronomer Christiaan Huygens made this, the first reasonably accurate sketch of Mars as seen through a telescope, in 1659. The V-shaped Syrtis Major is clearly seen.*

Contemporary with Huygens was the French astronomer Giovanni Domenico Cassini (1625–1712), who made observations from near Bologna, Italy, at the 1664 aphelic opposition of Mars, and again a couple of years later, recording the Syrtis Major as a dumbbell-shaped marking.

In 1666 Cassini accepted the invitation to become Director of Louis XIV's newly established Paris Observatory, where he was joined by Huygens. Both made observations of Mars from Paris in 1672, when the planet was at a perihelic opposition. Cassini made positional measurements, using Mars to help refine the size of the astronomical unit. Huygens continued to study the planet's markings, and is credited with discovery of the planet's polar caps.

After leaving Paris in the 1680s, Huygens, a skilled optician and clockmaker, continued to observe Mars with ever-larger, self-built aerial telescopes, including one with a 220 mm diameter object glass and a focal length of 64 meters! None of these giant instruments really revealed much more than Huygens' and Cassini's earlier telescopes, however, and it was to be some time before further advances in optics allowed more detailed examination of the Red Planet.

From the late 1680s onward, Cassini's nephew Giacomo Filippo Maraldi (1665–1729) observed Mars from Paris, making particularly noteworthy studies during the perihelic apparitions of 1704 and 1719. Maraldi noted the variable appearance of Mars' features, both from one apparition to the next and on a month-to-month basis. He thought the markings to be clouds, rather than surface features. By observing some of the more prominent and permanent markings, Maraldi found a rotation period for Mars of 24 hours 40 minutes (much in accordance with the value earlier found, independently, by Huygens and Cassini). Maraldi's 1719 observations documented the retreat of the south polar cap, which was then, as in 2003, presented toward Earth during Mars' southern summer.

The late 18th century

For several decades after Maraldi, Mars seems to have been neglected observationally. The next significant studies began with William Herschel (1738–1822), who was among the first to use large reflecting telescopes for astronomical observing. Herschel is famous for his discovery, in 1781, of the planet Uranus, from Bath, England, and for his later systematic cataloging of deep sky objects. In 1777 Herschel observed Mars' polar caps, and during the perihelic opposition of July 1781 he was able to refine the planet's rotation period to 24 hours 39.4 minutes (reasonably close to today's accepted value of 24 hours 37.4 minutes). On receipt of royal patronage, Herschel moved to Datchet, near Windsor, in 1783. In that year he was able to carry out extensive observations of Mars during its favorable presentation in September and October. Among his important findings were confirmation of Maraldi's observation that the south polar cap was not centered exactly at the pole, and a determination of Mars' axial tilt.

Herschel's drawings clearly showed the "Hourglass Sea" (a common early description of the Syrtis Major) and features identifiable as Sinus Sabaeus and Sinus Meridiani. His observations, made through a 470 mm aperture reflector, showed evidence for long-term changes in the surface features – some of the features shown in Maraldi's sketches are absent from those made by Herschel.

After 1783 Herschel concentrated on more distant targets, but the study of Mars was continued by the very capable German lunar and planetary observer Johann Hieronymus Schroeter (1745–1816). Schroeter's private observatory, at Lilienthal, near Bremen, was equipped with reflecting telescopes which used mirrors made by Herschel. From 1785 until 1802 Schroeter made observations, noting the variability of Mars' features (which, like Maraldi, he believed to be clouds), including the polar caps. Schroeter's observational records, significant parts of which were not published until 1887, provide a valuable insight to Mars' telescopic appearance in the late 18th century.

The early 19th century

With the development and use of compound lenses to counter the worst effects of chromatic aberration, the early 19th century saw a revival in the popularity of refractors for observational work. Most of the observers who made detailed studies of Mars in the 1800s and into the early 20th century did so using large-aperture refractors. With the use of compound lenses, these refractors could be kept to more manageable focal lengths than the cumbersome instruments used by, for example, Huygens 150 years earlier.

Two important observers of the first half of the 19th century were the Germans Wilhelm Beer (1797–1850) and Johann Heinrich Mädler (1794–1874). In 1828 Beer set up a private observatory equipped with a superb 95 mm refractor built by Fraunhofer, and the instrument was used by Beer and Mädler from 1830 onward to produce what was regarded for a great many years as the definitive map of the Moon. During the perihelic opposition of September 1830, Beer and Mädler turned their attention to Mars, arriving at the conclusion that the planet's dark markings were, in fact, permanent (rather than clouds, as had been suggested in the past).

Adopting a circular patch (which corresponds on more modern maps to the Sinus Meridiani) as their longitude zero, Beer and Mädler set about producing a map of Mars, which was eventually completed in 1840. While – for obvious reasons! – not as detailed as their lunar map, Beer and Mädler's first attempt at mapping Mars represented significant progress. Features on the map were identified simply by letters.

Mid-19th-century observers

The increased availability of better telescopes from the mid-19th century onward made detailed study of Mars accessible to more observers, many of them skilled draughtsmen. Among those whose careful renditions improved the developing picture of Mars' surface features were William Lassell (1799–1880) and Norman Lockyer (1836–1920). The Jesuit Father Angelo Secchi (1818–78) at the Collegio Romano made detailed observations at the perihelic opposition of 1860. He used a 240 mm refractor, which offered views magnified by up to ×400. An improved Mars map was prepared by Frederic Kaiser, Director of Leyden Observatory.

Another observer whose work led to still better maps was the Reverend William Rutter Dawes (1799–1868), who used, among other instruments, a 200 mm Cooke refractor at Haddenham, Buckinghamshire, UK, to examine Mars in the 1860s. Dawes' keen eyesight resolved the zero-longitude feature previously drawn as a circular or elongated patch into what became known as "Dawes' Forked Bay."

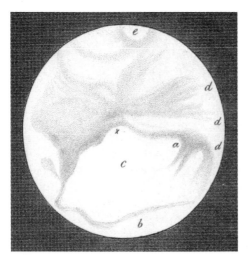

◄ William Dawes made this drawing using a 200 mm aperture refractor on November 20, 1864. Syrtis Major, Hellas and Sinus Sabaeus are all well shown, together with many streaky features. The "forked bay" (Sinus Meridiani), later named for Dawes, is close to the letter "a."

◀ *Giovanni Virginio Schiaparelli was one of the most distinguished students of Mars.*

Dawes' accurate drawings formed the basis for a map produced in 1867 by Richard A. Proctor (1837–88), the writer of numerous popular books on astronomy. Proctor's map was the first to introduce names for Martian features. The names were based on those of observers, and included the Lockyer Sea (modern Solis Planum), Mädler Continent (Chryse, Tharsis) and Kaiser Sea (Syrtis Major).

In September 1877 Mars came to a perihelic opposition against the stars of Aquarius at a distance of 56 million kilometers (35 million miles) – circumstances rather similar to those in 2003. During this return, American astronomer Asaph Hall (1829–1907) used the 660 mm Alvan Clark refractor at the US Naval Observatory in Washington to search for satellites. Hall found Phobos and Deimos over the nights of August 16 and 17.

Schiaparelli and the canals

The 1877 apparition also marked the beginning of observations by one of the greatest of all students of Mars, Giovanni Virginio Schiaparelli (1835–1910). He already had a distinguished history of astronomical achievement behind him – among his notable work was proving the link between the Perseid meteors and Comet Swift–Tuttle. Schiaparelli was attached to the observatory at the Brera Palace in Milan, where a 220 mm Merz refractor had been installed in 1874. Schiaparelli was to use this fine instrument over the next quarter-century, at magnifications of ×322, and sometimes ×462, to make a new map of Mars based on measurements made with a micrometer in the eyepiece.

While preparing his map, Schiaparelli found that the Proctor nomenclature did not cover the full range of features revealed by his acute eyesight and excellent optics. He began to adopt terrestrial names, calling bright features after lands (Hellas, for example, is

▼ *In this drawing by Giovanni Schiaparelli from the Brera Observatory in Milan, canali are shown as linear features in the deserts, running* *north from Sinus Sabaeus and Sinus Meridiani. The apparent junction where many of the canali meet was named Ismenius Lacus.*

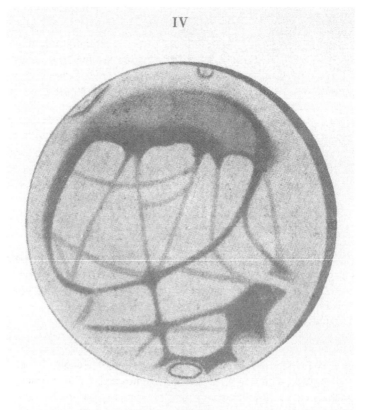

IV

1886 Aprile 5

$\omega = 350°$

Greece) while the dark areas were bodies of water (*maria*, or seas, which he held them to be). With only minor amendments, Schiaparelli's system became accepted as the standard until close-up spacecraft images forced a revision.

Among the features that Schiaparelli recorded – usually glimpsed during short-lived moments of good seeing – were narrow linear dark markings. Following the precedent of Secchi, about 20 years earlier, Schiaparelli referred to these in his native Italian as *canali*. Translated to English, *canali* can be taken either as "channel" or "canal:" adoption of the latter by some workers led to a most controversial period in the study of the Red Planet.

Schiaparelli continued his observations through the apparition of 1879 and beyond. In 1879 he detected the white cloud-enshrouded Nix Olympica (Olympus Mons in modern nomenclature). He also continued to record an ever-proliferating number of canals. Schiaparelli recorded some of these as undergoing doubling, describing the process as "gemination." Gemination, he thought, was seasonal, occurring during local spring. Over successive apparitions up to 1890, Schiaparelli took increasingly to drawing Mars as covered with networks of linear features.

Many others were sceptical of the canals' reality: Asaph Hall, for example, never recorded them, and observers at the Lick Observatory (including James Keeler, who detected the gap, now called the Encke Division, in Saturn's outer ring) were likewise unconvinced, finding only a few linear features, none of which corresponded particularly well with anything suggested by Schiaparelli.

Observers at Nice Observatory, in France, however, were more enthusiastic in following Schiaparelli's lead in drawing canals. The great French popularizer of astronomy Camille Flammarion (1842–1925) was convinced of the canals' reality. William Henry Pickering (1858–1938), Director for a time of the Arequipa Observatory high in the Peruvian Andes, was certain that the canals existed; he also adhered to the view that the greenish tint in Mars' dark features was the result of vegetation.

The canal controversy

Pickering was to play a part in the work of Percival Lowell (1855–1916), who was surely the most controversial of all characters to study Mars. Lowell came from a Boston, Massachusetts, family which had made its fortune in the textile industry. Educated at Harvard, he was well traveled, spending many years in the Far East. Lowell's interest in Mars began in the early 1890s, and when he returned finally to the United States he set about establishing an obser-

◄ *Percival Lowell was a major proponent of canals on Mars being proof of intelligent life on the planet.*

vatory at Flagstaff, Arizona, for the express purpose of studying the planet. Founded in 1894, the Lowell Observatory was sited at an altitude of 2100 meters (6900 ft), to take advantage of favorable seeing conditions, and was initially equipped with refractors of 300 mm and 460 mm aperture. As assistants, Lowell employed Pickering and another Harvard astronomer, Andrew Ellicott Douglass (1867–1962).

Lowell and his assistants began their study of Mars in 1894, making drawings that showed a more intricate lacework of canals than even those of Schiaparelli. Lowell was convinced of the reality of the canals, holding it "self-evident" that they were "the work of some sort of intelligent beings." In Lowell's view, the canals were an irrigation system used by a Martian civilization to combat the advanced drying state of their world. The canal network carried meltwater from the poles to the desert regions.

Lowell was a relentless self-publicist: his views were published in countless articles and in a book simply entitled *Mars* (1895). His ideas caught the public imagination, but were less well received by astronomers, many of whom were offended by Lowell's aggressive style: he was dismissive of those who did not see his canals, for example, suggesting that their failure to do so came down to poor equipment or inadequate seeing at their observing sites.

Observations continued in 1896, when Lowell took delivery of a still-larger telescope, a 610 mm refractor. This he used both at Flagstaff and in Mexico. At this time, Lowell also turned his attention to Mercury and Venus, drawing the latter as having a system of radial "spokes."

From 1897 ill health caused Lowell to take a four-year break from his work. His assistant Douglass, however, carried on with observations. During this hiatus, opposition to the canals drawn by Lowell, and to his linear features on Venus, was growing among astronomers. Some experienced planetary observers, including Edward W. Maunder (a founder of the British Astronomical Association), and even Douglass, carried out experiments using artificial planet disks viewed over long distances through telescopes. Their findings began to point strongly to the linear canals being a trick of the eye and mind, which has a tendency to connect closely spaced points with illusory lines.

Using the 910 mm (36 inch) Lick refractor, Edward Emerson Barnard (1857–1923), one of the most highly skilled visual observers of the day, found no trace of the canals, only a few streaky details. Others concurred that under the best seeing conditions any apparent linear structures usually broke up into smaller points. To many observers, the canals were simply an illusion produced by fine detail at the limits of resolution imposed by atmospheric conditions. Another blow to the theory of canals came in 1894, when spectroscopic studies of Mars made at Lick failed to detect water.

The appearance of dark spokes on Venus was rejected outright by contemporary observers. Recently, William Sheehan and Tom Dobbins have suggested, quite plausibly, that what Lowell recorded was actually a reflection in the eyepiece of the blood vessels in the back of his eye!

12-inch diaph. λ 268°

power 490 seeing 6

◄ This is one of Lowell's many Mars drawings showing canals in the area north of Syrtis Major. It was made on April 7, 1905, with a 600 mm refractor at the Lowell Observatory.

► *Eugène Antoniadi was an extremely skilled observer, whose studies using the Grand Lunette refractor at Meudon Observatory finally laid to rest the idea of a Martian canal network. This portrait, published in the British Chess Magazine for September 1907, was recently discovered by Richard McKim, and has never previously been published in an astronomical book or journal.*

Schiaparelli, whose initial observations of the canals triggered the controversy, maintained something of a distance from Lowell's activities, and, without completely rejecting the idea, he refused to commit himself to the opinion that Mars was inhabited. Failing eyesight led Schiaparelli to give up observational work in 1898, but he continued to take an interest in the study of Mars, an area where his authority was widely recognized.

Lowell returned to the fray in 1901 and 1903, making observations at Flagstaff. Douglass' doubts over the Venusian spokes had cost him his job, but Lowell had several new assistants. One of them, Carl Otto Lampland, obtained photographs which Lowell purported to show some of the canals; others were less convinced.

While Lowell's ideas met with widespread scepticism among scientists, they continued to have great popular appeal. Lectures by Lowell attracted large audiences, and his books *Mars and its Canals* (1906) and *Mars as the Abode of Life* (1908) were both well received by the public.

The tide of opinion regarding the reality or otherwise of the canals was turned following the September 1909 perihelic opposition. Significant observations were made at this return by Eugène Michael Antoniadi (1870–1944), using the "Grand Lunette" 830 mm refractor at the Meudon Observatory near Paris.

A skilled observer and draughtsman, Antoniadi was employed at Flammarion's private observatory at Juvisy from 1893 until 1902. As

Director of the BAA Mars Section between 1896 and 1917, he was responsible for compiling detailed reports after each apparition. Initially, Antoniadi was happy to accept the existence of canals, though perhaps not in the numbers that Lowell and his colleagues suggested. In time, however, he became less certain, tending toward the view that the canals might be a product of straining to see details at the limit of detection.

Given the opportunity to observe with the great Meudon refractor in 1909, Antoniadi enjoyed some exceptional views of Mars. On September 20, with Mars showing a 24-arcsecond disk among the stars of Pisces, the planet's surface appeared to have a great wealth of fine detail – mottling, spots and stippling – some of it almost too intricate to draw. On the best nights, when the seeing was very steady, there was no hint of "geometry," or linear canal-like features. For most people, Antoniadi's observations were sufficient proof that Mars, while possessing a fantastically detailed surface, was bereft of the multitudinous canals drawn by Lowell and his followers. The canals were nothing more than an illusion brought about by the limitations of viewing fine structure on a distant globe with a small apparent diameter. The apparent doubling phenomenon – Schiaparelli's gemination – could likewise be put down to eye strain and slight errors in focusing.

▼ Below left is a drawing made by Eugène Antoniadi on February 8, 1929, using the great Meudon refractor. Solis Planum is complex and, unusually, extended to the northwest. Below right is an Antoniadi drawing from October 6, 1909, when Mars was close to a very favorable, perihelic opposition. It was made with the Meudon 830 mm refractor. Solis Planum is near the middle of the disk, with Mare Sirenum well on view.

While the canals fell from widespread acceptance, Lowell still adhered to their reality, and continued his observations at Flagstaff through the increasingly unfavorable apparitions of 1911, 1913–14 and 1916. By this time, he had installed a 1.02 meter reflector.

Lowell died in November 1916, still convinced of the existence of intelligent life on Mars. While some continued to support the possibility of canals on Mars, even up to the dawn of the Space Age in the 1960s, the idea had largely fallen from favor by the 1920s even in popular thought.

The rise of amateur organizations like the British Astronomical Association (BAA) and the Association of Lunar and Planetary Observers (ALPO) in the 20th century led to increased numbers of observers studying Mars – and the other planets – on a regular and frequent basis through relatively small instruments. In many ways, this represented a "golden age" for visual observation and had a number of benefits. Amateur observations of the development and spread of yellow clouds (dust storms), for example, have been valuable in learning more about Martian meteorology. Both amateur and professional studies have led to improved maps of Mars' albedo features. By the 1950s, however, many felt that exploration of Mars by telescope had been taken as far as it could be in many respects. The development of interplanetary spacecraft would, from the 1960s, lead to some new and unexpected discoveries.

SPACE AGE MARS

Alongside the development of the lunar landers, exploration of the other planets – principally Mars and Venus – was an obvious target for early spacecraft missions in the 1960s. Sending a probe to Mars is a considerable challenge. The fuel and energy requirements for a direct flight are prohibitive. Instead, the most economical route is via what is called a Hohmann transfer orbit, whereby the craft is launched into an orbit around the Sun which intercepts that of Mars. Such an orbit is achieved by launching the spacecraft during the period when Mars is 45 degrees ahead of Earth in its orbit. Typically, an interplanetary cruise of about 260 days is required to reach the Red Planet.

In each Martian year of 687 days, there is only a brief interval – the "launch window" – during which the planet is favorably positioned for a probe to be sent. Some launch windows, such as 1971, 1988 or 2003, are more favorable than others, since the Earth–Mars distance is smaller and less fuel is required. Reducing the fuel load increases the amount of instrumentation (sensors, cameras and so on) that a probe can carry.

The usual mission profile involves launch to Earth orbit, followed by a less energy-demanding rocket burn from "parking" orbit into a Mars-intercept trajectory. Small corrections usually have to be carried out using onboard thrusters while the probe is en route.

Launch windows to Mars open every 26 months, and several missions are planned to exploit these windows in the coming years. The long flight time is, of course, a serious barrier to manned Mars missions, and for the foreseeable future robotic explorers will continue to be the preferred means for gathering detailed information about the planet.

Early missions

Having managed to send spacecraft past, then around, the Moon, the Russians made several attempts to reach Mars in the early 1960s. Their 1962 Mars 1 probe – essentially identical to the lunar spacecraft – was well on its way to the Red Planet when contact was lost.

NASA, in the United States, enjoyed better results: following the launch failure of Mariner 3, the sister craft, Mariner 4, was launched on November 28, 1964, arriving at Mars on July 5 the following year. Its single television camera returned 21 images, showing – to scientists' surprise – cratered terrain in Mars' southern hemisphere. Measurements from Mariner 4's instruments showed the Martian atmosphere to be extremely thin, and surface temperatures very low.

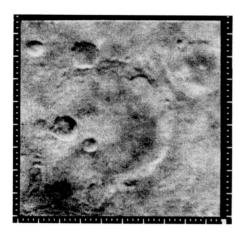

◄ *Scientists were surprised when the earliest Mars spacecraft missions returned views, like this one from Mariner 4 in 1965, showing cratered terrain in the planet's southern hemisphere.*

Little or no magnetic field was detected. Mariner 4 flew past Mars at a minimum distance of 9600 km (6000 miles).

Somewhat overshadowed at the time by the Apollo 11 lunar landing, NASA's next two missions, Mariner 6 and Mariner 7, were launched in 1969, making their fly-bys of Mars on July 31 and August 5, respectively. Mariner 6 flew 3400 km (2100 miles) above the Meridiani Terra region, imaging still more cratered terrain and confirming the Martian atmosphere to be largely composed of carbon dioxide. Mariner 7 flew similarly close, returning over 100 images, including views that showed the Hellas basin in detail.

The Russians tried again for the Red Planet in 1971, with the probes Mars 2 and Mars 3. Each comprised an orbiter together with a lander section designed to make a soft touchdown and send back images from the surface. On November 27, 1971, the Mars 2 lander crashed at a site in Hellas, becoming the first man-made object on the planet. Mars 3 arrived on December 3, its lander apparently making a safe descent in the southern hemisphere's Mare Sirenum region, transmitting briefly before communications were lost.

Mariner 9

Without doubt, the most successful of the 1971 missions was NASA's Mariner 9 orbiter, which gave a whole new view of Mars. The previous fly-by missions had given only glimpses of cratered regions, mostly in the southern hemisphere. From its elliptical 12-hour orbit, with a low point 1400 km (870 miles) above Mars and a high point of 14,000 km (8700 miles), Mariner 9 revealed a world with complex geological structures, including the giant Tharsis volcanoes, the Valles Marineris canyon system and, perhaps most exciting, obvious water flow features.

Mariner 9 was originally conceived as one of a pair of orbiters, but the sister craft – Mariner 8 – was lost during launch in early May 1971. The mission was modified, and Mariner 9 was successfully launched on May 30, 1971, entering orbit around Mars on November 14. Mariner 9 was equipped with instruments to study Mars' surface temperature and atmosphere. More importantly, it carried cameras that were capable of resolving features down to one kilometer in size.

Like other Mars probes, Mariner 9 was powered by solar cells. Sunlight is still sufficiently strong at the distance of Mars' orbit to produce enough electricity. (The deep space Pioneers and Voyagers, by contrast, had to rely on their onboard radioisotope thermonuclear generators.)

Mariner 9's arrival at Mars coincided with the height of the great global dust storm (attendant adverse surface conditions may account for the failures of the Mars 2 and 3 landers a couple of weeks later). Just as it prevented Earth-based observers from seeing much, the dust also obscured the Martian surface from Mariner's cameras; only the summits of the Tharsis volcanoes, poking through the shroud, were visible, and it was not until January 1972 that the main mapping program could commence.

In the meantime, in late November 1971, Mariner 9 obtained the first close-up images of Phobos and Deimos. These were shown to be small, heavily cratered bodies, not inconsistent with their presumed nature as captured asteroids.

▼ *Mariner 9 was the first spacecraft to be placed in orbit around another planet. It entered orbit around Mars in 1971, giving* *scientists the first really detailed views of the whole planet. Previous fly-by missions had only imaged small parts of Mars.*

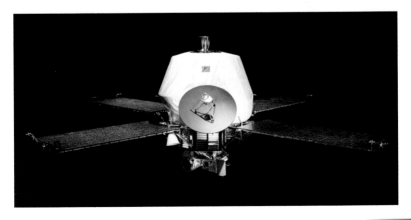

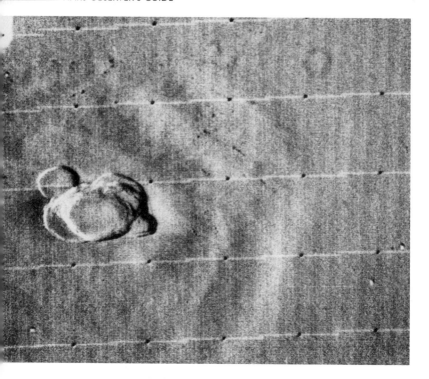

▲ *One of Mariner 9's first images, this photograph shows the summit crater of* Olympus Mons reaching above the 1971 planet-wide dust storm.

When the dust finally cleared, Mariner 9 began to record spectacular images of hitherto unsuspected features. It was this mission that brought to light the difference between the two Martian hemispheres – the ancient, cratered south and the more recently modified plains-dominated north – and revealed the truly enormous scale of the shield volcanoes, including Olympus Mons. Sand dunes, dried water channels, craters and canyons were all recorded in detail.

Mariner 9 operated until it ran out of fuel for its attitude-control thrusters in late October 1972. The inert spacecraft remains in orbit around Mars today.

Beyond Mariner

Budget constraints held back the next NASA missions to Mars – the Vikings – until 1975–6. The Russians, however, attempted to make full use of the July/August 1973 launch window to send four spacecraft to Mars. Mars 4 and Mars 5 were intended as orbiters, while

Mars 6 and Mars 7 were to be fly-by vehicles releasing landers. The landers were to relay their data from the surface via Mars 4 and 5. Engine failure resulted in Mars 4 cruising past the planet – though it did take some images – on February 10, 1974. Mars 5 was successfully placed in orbit two days later. Images from these two craft compared favorably with those from Mariner 9. The Mars 6 lander touched down in the Mare Erythraeum region in the southern hemisphere on March 12, 1974, but its signal was lost during the descent. Mars 7, meanwhile, had missed the planet on March 9.

The Vikings

The two Viking craft each consisted of an orbiter, heavily based on the Mariner 9 design, and a lander. Viking 1 was launched on August 20, 1975, Viking 2 a few weeks later on September 9. The spacecraft were placed in orbit around Mars on June 19 and August 17, 1976, respectively. Like Mariner 9, the Vikings had highly elliptical orbits, with closest approach around 1500 km (930 miles) above the Martian surface, and greatest distance 33,000 km (21,000 miles).

The landers carried miniature automated laboratories, whose purpose was to search for signs of possible life in the Martian soil. Potential landing sites for such investigations – preferably flat for a safe landing, and associated with water flow features – were identified from Mariner 9 images. On arrival, more detailed reconnaissance was carried out using the orbiters' cameras, which had a finer (typically 300 meters/1000 feet) resolution than those on Mariner 9.

The Viking 1 lander touched down successfully on Chryse Planitia on July 20, 1976, while the Viking 2 lander came down on Utopia Planitia, 6000 km (3700 miles) away, on September 3. Following release from its orbiter, each lander underwent deceleration in the

▶ *Viking 1 and 2 placed landers on the Martian surface in 1976. They sampled the soil and looked for evidence of life, returning meteorological data and images over long periods of operation. Shown here is a model of a Viking lander.*

◄ This view from the Viking 2 lander on Utopia Planitia shows a thin coating of frost on the ground just after local sunrise on May 18, 1979.

Martian atmosphere, using a protective aeroshell/heat shield, before descending on parachutes, and finally making a soft touchdown cushioned by retrorockets. Cameras on the landers recorded striking panoramas, dominated by boulders scattered across a gently rolling, dusty landscape. Fine dust suspended in the atmosphere gave the Martian sky a salmon-pink hue.

Instruments on the landers recorded local meteorology in detail, showing wide temperature variations between day and night: the temperature at the Chryse Planitia site, for example, ranged from −33°C to −86°C (−27°F to −123°F). Wind speeds of around 30 km/h (20 mph) were recorded, with some stronger gusts. Seismometer measurements recorded a weak, distant tremor, the signal from which suggested the presence of frost deposits in the soil.

The soil itself was directly sampled using a robotic arm and scoop on each lander. Instruments in the onboard laboratories included an X-ray spectrometer for analysis of the samples. The large blocks on the surface were found to be basalts – volcanic rock of local origin broken up by impact processes. The fine-grained dusty soil is a product of rock weathering.

Of greatest interest to many, of course, were the biological experiments, which gave somewhat inconclusive results. The first impression, certainly, was that the Martian soil contained little or no organic material. As we shall see in Chapter 10, the debate over the presence or otherwise of life on Mars continues.

The Viking 2 lander functioned until April 12, 1980, and the Viking 1 until November 13, 1982. During their functional lifetimes, both landers returned daily weather information, and a total of almost 10,000 images came back from the Martian surface.

The Viking orbiters mapped 97% of Mars to a resolution of 300 meters (1000 feet), and some parts to as small as 25 meters (80 feet), in some 51,000 images. In May 1977 the Viking 2 orbiter approached to within 28 km (17 miles) of Deimos and was able to measure its density: at 2 g/cm^3, it is comparable to that of carbonaceous chondrite meteorites, a result which is in keeping with the presumed origin of the Martian satellites as gravitationally captured asteroids. The Viking 2 orbiter operated until July 1978, being outlasted – until August 1980 – by that of Viking 1.

Back to Mars

After the hugely successful Viking missions, there was a long gap before further spacecraft exploration of Mars was attempted. The Russians launched two probes, Phobos 1 and 2, during the very favorable launch window in July 1988. They planned to place a lander on Mars' moon Phobos, but the mission failed. Contact with Phobos 1 was lost while it was en route to Mars, and Phobos 2 was lost in March 1989 when it was beginning to close in on its target.

Similar bad luck befell NASA's Mars Observer, which was launched in September 1992. Days before its final approach to Mars, in August 1993, the probe was lost, possibly to a catastrophic explosion in a propellant tank.

Mars 96, an orbiter developed from the earlier Phobos probes, made it to Earth orbit on November 16, 1996. From here, a second

▶ The Mars Pathfinder and the Sojourner rover are shown on the Martian surface in this artist's impression.

rocket burn was supposed to send it on a trajectory toward Mars, but instead it re-entered Earth's atmosphere and was destroyed the following day, thus continuing Russia's string of misfortune in the study of Mars. (Russian missions to Venus, by contrast, have mostly been great successes.)

Mars Pathfinder

The new phase of Mars exploration got underway with the very successful NASA Mars Pathfinder mission. This comprised a lander with a small, separate rover vehicle capable of moving around on the surface. Mars Pathfinder was launched on December 4, 1996, and reached the planet on July 4, 1997. Following aeroshell braking and parachute descent, the 264-kg Pathfinder lander was dropped in a cocoon of air balloons, hitting the surface at a velocity of 18 m/s (60 ft/s). It bounced – up to a height of 15 meters (50 feet) – across the surface for a distance of about a kilometer before coming to rest a couple of minutes later. The airbags were shed, and the "petal" solar panels unfolded, putting the lander into operation at a site in the Ares Vallis, in the Chryse Planitia, approximately 800 km (500 miles) east-southeast of the Viking 1 landing site.

Ares Vallis (9°42′N, 23°24′W) was chosen for the landing site because it is a floodplain on which water from catastrophic outflows (with a volume perhaps equivalent to the Great Lakes combined) converged in the distant past. Project scientists felt that this location might offer samples of material, carried by the flood, from several places. This proved to be the case, with rocks at the landing site oriented as might be expected had they been swept along by floodwater. The local terrain seemed broadly similar to the Viking landing sites, with perhaps more rocks.

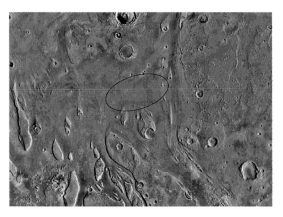

◀ The Mars Pathfinder landing site is enclosed by the ellipse in this orbiter image of the Ares Vallis region. The site, near the Viking 1 lander location on Chryse Planitia, is in a region of water outflow features.

▲ *A panoramic view from the Mars Pathfinder lander shows the rocky* *terrain at Ares Vallis. The Sojourner rover can be seen at center right.*

On July 5 the lander was renamed as the Carl Sagan Memorial Station in honor of the late American scientist and popularizer of science who had worked on past missions, including the Mariners and Vikings. The lander deployed the small (630 mm by 480 mm, 260 mm high; 25 by 19 by 10 inches) Sojourner rover to analyze several rocks in the vicinity. This six-wheeled rover descended the ramp on to the surface of Mars on July 6.

Sojourner carried cameras to image the rocks, and an X-ray spectrometer to undertake chemical analysis. Many of the rocks at the Ares Vallis site were found to be conglomerates, which is consistent with a sedimentary origin, perhaps on the floor of a lake or other body of standing water.

Powered by solar panels (unlike the Viking landers, which had small radioactive power sources), the Pathfinder mission ran for 83 sols (Martian days), three times longer than originally anticipated. During the mission, the lander returned 16,000 images and Sojourner returned 550. Particular achievements were the detailed analysis of rocks and meteorological data. The mission finally ended on September 27, 1997, once power became depleted.

Mars Global Surveyor

On November 7, 1996, NASA's Mars Global Surveyor (MGS) was launched on a slower trajectory to Mars. Following its orbital insertion on September 12, 1997, it greatly advanced the detailed mapping of the planet's surface. Initially, MGS was in a highly elliptical orbit, 260 by 54,000 km (160 by 34,000 miles). This was slowly altered by the use of thrusters and a daring aerobraking technique, which employed the drag of the tenuous outer Martian atmosphere on the spacecraft to slow it down.

Mars Global Surveyor reached its required circular mapping orbit by February 1999. Now an average of 378 km (235 miles) above the surface, MGS carried out its mapping in a polar orbit perpendicular to the planet's equator, passing south to north. The orbit, described as Sun-synchronous, placed MGS above the afternoon part of the planet during it imaging runs. Coupled with Mars' rotation, this two-hour orbit allowed a complete mapping cycle every 26 days. MGS carried a camera system that was capable of resolving features down to one meter in size.

▼ *Mars Global Surveyor entered orbit around the planet in 1997. Over the next five years, it mapped the surface to a resolution as small as one meter.*

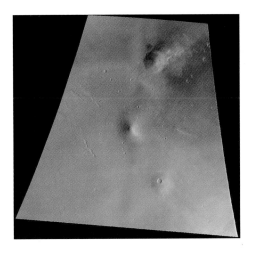

◀ The Elysium volcanoes
are well shown in this
Mars Global Surveyor view.

Among the major findings from Mars Global Surveyor has been confirmation of the presence of substantial quantities of water locked up in permafrost deposits in the Martian soil, particularly at high latitudes in the planet's southern hemisphere. This finding came from the detection of hydrogen using the orbiter's gamma ray spectrometer. An identical instrument aboard the Lunar Prospector spacecraft provided compelling evidence for water in the permanently shadowed craters of the Moon's poles in 1998. Mars' permafrost, in a layer less than a meter below the surface, is sufficient in quantity to allow for the past existence of lakes in which sedimentary rocks could have formed. MGS continued its mapping program until April 2002, its work having spanned more than a Martian year (687 days). The orbiter will serve as a relay for missions in 2003–4.

Losses and recovery

The great success of the Mars Pathfinder mission was followed by a couple of disappointing failures. NASA's Mars Climate Orbiter was launched on December 11, 1998. It arrived at Mars late the following year when, thanks to a calamitous mix-up over the use of imperial and scientific units in setting its trajectory, the spacecraft burned up in the planet's atmosphere. Launched on January 3, 1999, the Mars Polar Lander vehicle reached Mars on December 3 the same year, but crashed – as a result, it is thought, of premature shutdown of its retrorockets. The Mars Polar Lander was meant to deliver two penetrators to investigate the Martian subsurface, but these, too, were lost.

▲ *Mars Odyssey, shown above the planet in an artist's impression, has* *achieved many of the scientific goals of the ill-fated Mars Climate Orbiter.*

Much of the science that should have been carried out by Mars Climate Orbiter was achieved by a relatively cheap replacement. Launched on April 7, 2001, Mars Odyssey reached the Red Planet on October 24 the same year. It was placed in a circular polar orbit at an altitude of 400 km (250 miles), following aerobraking maneuvers lasting 76 days. Originally to have been called Mars Surveyor 2001, the spacecraft was renamed in honor of the British science fiction author Arthur C. Clarke and his famous novel *2001: a Space Odyssey*. Mars Odyssey is carrying out global observations of Martian topography and meteorology, and is mapping the near-planet radiation environment. Spectacular images, to a resolution of 100 meters (330 feet), have been returned from the orbiter's THEMIS (Thermal Emission Imaging System) instrument. Mars Odyssey is also equipped with a neutron spectrometer, which confirmed the Mars Global Surveyor finding that the planet has a substantial reservoir of water in its surface permafrost.

If all goes to plan, Mars Odyssey will serve as an orbital relay station for missions to be sent to the planet during the 2003 launch window, and will continue its own mapping program until July 2004.

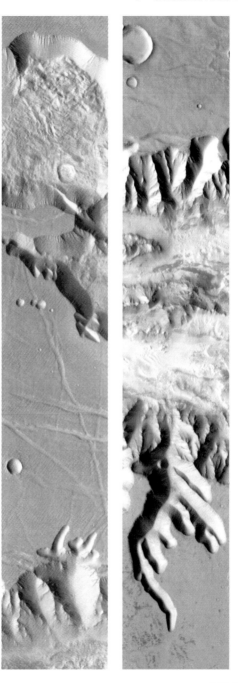

▶ The THEMIS instrument aboard Mars Odyssey has returned some stunningly detailed images of the surface. These images show parts of the Valles Marineris complex. THEMIS takes thin mapping strips.

Missions for 2003–4

NASA plans to send two identical Mars Exploration Rovers to the Red Planet in 2003. With projected launch dates of May 22 and June 27, these should land on Mars on January 2 and January 20, 2004. Possible landing sites have been identified in Melas Chasma (in the Valles Marineris), Meridiani Terra, Gusev crater (a large, ancient impact crater in the southern highlands, which may once have held a lake) and Athabasca Vallis (in the Elysium region). In the course of a 90-day mission, each rover is expected to cover 100 meters (330 feet) of ground per day, taking images and sampling rock surfaces freshly exposed using their diamond-tipped rock abrasion tool ("RAT"). The Mars Exploration Rovers will be cushioned on landing by an airbag system similar to that used for the 1997 Mars Pathfinder landing. Weighing 180 kg, the Mars Exploration Rovers will, like Pathfinder, be powered by solar panels.

The European Space Agency, Italian space agency and NASA are collaborating on the Mars Express mission, which is scheduled for launch in June 2003 aboard a Soyuz Fregat rocket from Baikonur in Kazakhstan. Mars Express will examine the Martian surface and atmosphere from a polar orbit, and will also search for subsurface

▼ Among the missions planned for 2003–4 are NASA's twin Mars *Exploration Rovers, which will cover large tracts of the Martian surface.*

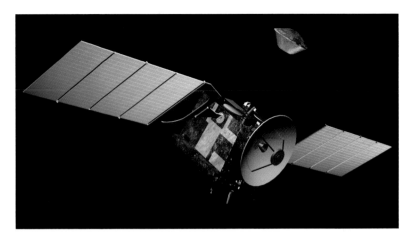

▲ *The European Space Agency's Mars Express spacecraft will study the Martian surface and atmosphere from a polar orbit.*

In this artist's impression, Mars Express is seen releasing the Beagle 2 lander shortly before arriving at the Red Planet.

water. A stereo camera system will resolve features in three dimensions to a scale of 10–30 meters (30–100 feet), and sometimes to as little as two meters. Many of the orbiter's instruments are identical to European experiments lost in the Mars 96 failure. Mars Express should reach the planet late in 2003.

Mars Express will deliver the United Kingdom's Beagle 2 lander. A spring mechanism will eject Beagle 2 during the final approach to Mars, and the planet's gravity will pull the lander in. A heat shield will protect the lander during entry into Mars' atmosphere, and once it has slowed to 1600 km/h (1000 mph), parachutes will be deployed to slow it further. The final descent will be cushioned by airbags. All being well, Beagle 2 will bounce to its final resting place on the Isidis Planitia (10°36′N, 270°W).

Isidis Planitia, near Syrtis Major, has been selected as the landing site because it is believed to be a sedimentary basin, filled in the distant past by water – perhaps enhancing the possibility that evidence for life might be preserved there.

On release of the protective airbags, Beagle 2's solar panels will be deployed, followed by the robotic arm, which carries various experiments on its end. The first priority will be to take images of the landing site, and then rock and soil samples will be collected for analysis. Among the instruments on the lander are a mass spectrometer: Beagle 2 will look for organic material as evidence for the past or current presence of life on Mars. Where past missions have largely sampled the topsoil, Beagle 2 will be equipped with "moles" capable of burrowing under the surface to collect material.

Beagle 2 takes its name from HMS Beagle, the ship aboard which Charles Darwin sailed on his 19th-century voyage of discovery, during which he gathered the evidence to put his theory of evolution on a firm footing. The Beagle 2 team, led by Professor Colin Pillinger of the UK's Open University, hope that their rather smaller craft will provide better insights to whether or not Mars is, or has been, the abode of life.

Beyond 2004

Already on its way to Mars is Japan's Nozomi ("Hope") spacecraft, which was launched back on July 3, 1998. Nozomi was placed on a long trajectory, dependent on gravitational boosts ("slingshots") from close passages to the Moon and Earth. During the April 2002 Earth fly-by, Nozomi was affected by solar flare activity, and there

▼ This artist's impression shows Beagle 2 on the Martian surface with its solar panels deployed and an array of instruments and experiments on its robotic arm.

are concerns that damage to its systems may endanger the mission. After a second gravity boost from Earth in June 2003, Nozomi is due to enter orbit around Mars in June 2004, from where it will study the planet's upper atmosphere and image its surface. Images of Phobos and Deimos will also be taken, and Nozomi is expected to make improved measurements of Mars' weak magnetic field.

During the August 2005 launch window, NASA plans to send the Mars Reconnaissance Orbiter to image the planet's surface at better than one-meter resolution. It will also look for evidence of water. Aerobraking will place the craft in a 250 by 320 km (160 by 200 miles) polar orbit.

Future plans are likely to be still more ambitious: among the plans on the drawing board for the late 2007 launch window are, for example, multiple landers that will take simultaneous measurements from separate locations. Up to 2003, all the landers have investigated sites in the Mars' northern hemisphere, and scientists are keen to improve the variety of the sampling.

A future mission is also expected to return samples to Earth for examination in more rigorous laboratory conditions. At current estimates, costs and technical difficulties place the likely timing of such a mission beyond 2010. A manned Mars mission is still a long way off, but the renewed interest of the various national space agencies will certainly ensure that the planet is well mapped before it is visited by human beings.

Spacecraft have given a detailed picture of Mars beyond the wildest imagination of the early telescopic observers. Lowell's canals are long gone from any hope of reality, but in their place we have a world of fantastic natural landscapes – canyons, volcanoes, plains scattered with boulders, sand dunes and more. The lander and orbiter missions of 2003–4 will be avidly followed by many of us. When Mars Pathfinder was active on the surface during July 1997, the live-camera website established by NASA received 25 million "hits" per day, showing just how high is the level of public interest in the Red Planet. As Mars retreats from favorable Earthbound view at the end of 2003, planet-watchers should be able to keep up with developments via the Internet and other media links. We live in an exciting and fascinating time in the exploration of Mars.

LIFE ON MARS?

As we saw in Chapter 8, the idea that Mars might be the home of simple life is not new. While Lowell's canals have long been dismissed as illusory, and not the works of an advanced civilization, speculation that the spring wave of darkening (see chapter 4) might represent simple plant growth could not be dismissed until later in the 20th century, when spacecraft and other measurements showed the planet to be too dry to support it. Nevertheless, some scientists still adhere to the possibility that microbial life might have developed on Mars in the past, and could still be extant in the soil. Proving the existence or otherwise of Martian life remains an issue for the current generation of robotic explorers.

Life on Earth – and elsewhere?

Evidence from fossil bacteria shows that life arose on Earth quite soon after the planet's formation – certainly during the first billion years following Earth's condensation from the presolar nebula. Exactly how life originated is still uncertain. Probably the most popular theory with biologists is that chemical reactions, driven by lightning discharges and ultraviolet radiation from the Sun, led to the formation of complex molecules in the primitive atmosphere and oceans of Earth, which was then an oxygen-free environment. Organic (carbon-containing) molecules collected in the oceans, and where pools of water were trapped and evaporated, these molecules could undergo further reactions, leading eventually to self-replicating forms.

Laboratory experiments provide some good evidence in support of such a scenario. By sparking electricity through a gas mixture containing methane (CH_4), ammonia (NH_3), hydrogen (H_2) and water (H_2O), Stanley L. Miller and Harold C. Urey at the University of Chicago were able, in 1953, to collect, after a week, a sticky, brown tar-like residue containing several biologically important molecules, including amino acids, such as leucine, serine, threonine, lysine and so forth, and precursors of those involved in building up nucleic acids (the information-carrying DNA and RNA).

The emergence of life within the first billion years after Earth's formation is often cited as an argument against life having had a purely terrestrial origin: some prefer the idea, known as panspermia, of it having been delivered preformed from elsewhere by comets or meteorites. Their common assertion is that a timescale of a billion years is too short, given the required series of chemical reactions. However, the production of so many complex biological precursor molecules in an experiment lasting only one week must surely be an effective counter to this argument – there are a great many weeks in a billion years!

Astronomers studying nebulae in regions like Orion – in which stars and, we can fairly safely assume, planets are forming – have found the spectral signatures of many biologically important organic molecules. These signatures are also found in comets and in primitive carbonaceous chondrite meteorites, both of which preserve material largely unaltered since the time of the Solar System's formation. The organic molecules in comets and carbonaceous chondrites were present in the nebula from which the Sun and planets condensed 4.6 billion years ago, and it is not unreasonable to suggest that some of the prebiotic chemistry of Earth was supplemented by material delivered in comet and meteorite impacts. Most biologists would, however, baulk at the ideas of Sir Fred Hoyle and Chandra Wickramasinghe, who suggested that comets deliver clouds of influenza virus at the current epoch!

For most of its history, life on Earth has been very primitive. The first 2.5 billion years or so were dominated by bacteria. Life only began to get complex about a billion years ago, with the development of oxygen-liberating photosynthesis by blue-green algae. The production of oxygen changed the atmospheric composition, allowing the development of the stratospheric ozone layer (the action of solar ultraviolet on oxygen produces ozone, O_3, a triple-atom form of oxygen). This layer provided a screen from ultraviolet radiation, allowing colonization of the land and evolution of more advanced organisms, including ourselves.

There is no reason to suggest that Earth is the only planet in the Universe where such a chain of events might have been played out. Given the abundance of organic molecules in interstellar space, it is probable that similar chemical processes could have occurred, or are occurring, elsewhere. The burgeoning field of astrobiology investigates the possibility of life in planetary systems beyond our own.

More relevant with respect to Mars is the question of whether life might have arisen there during its earlier history. Like Earth, Mars would have been bombarded early on by organic molecule-rich comets and meteorites, and on a warmer, wetter early Mars, the chemistry of life may have had favorable conditions to get started. If it did, life clearly took a different evolutionary path – for instance, there is no great abundance of oxygen or a protective ozone layer in Mars' high atmosphere. This lack does not in itself, however, preclude the possibility of primitive bacterial life in the Martian soil. Establishing whether such life might exist was an obvious goal for the 1976 Viking landers.

The Viking experiments

The Viking landers' onboard laboratories carried three key experiments to look for evidence of life on Mars, testing soil samples collected by the scoop on the end of their robotic arms. The pyrolytic release experiment looked for assimilation of radioactive carbon (^{14}C) from carbon dioxide and carbon monoxide. The gas exchange experiment looked for metabolic by-products, such as carbon dioxide, following addition of nutrient medium. Finally, the labeled release experiment provided nutrient broth tagged with ^{14}C, seeking its incorporation into metabolic by-products.

Copious amounts of gas were indeed produced in the gas exchange experiment when a nutrient medium was added to Martian soil – an initially encouraging result for those hoping to find life. Most of the released gas was oxygen (O_2), with smaller amounts of carbon dioxide (CO_2). An experiment is, however, only as good as its control. In this case, the control involved heating the soil sample to 160°C (320°F), at which temperature even the most robust of terrestrial bacteria would be killed. This sterilized soil sample also released a burst of oxygen when incubated with nutrient medium, indicating that the reaction had nothing to do with biology, and everything to do with Martian soil chemistry.

The pyrolytic release experiment was carried out without water and nutrients. It looked for evidence of assimilation of ^{14}C-labeled CO_2 and carbon monoxide (CO) by any organisms that might have been

▼ The series of images shows sample collection by the Viking 2 lander's robotic arm. Small samples of soil – typically less than a cubic centimeter in volume – were analyzed in the Viking landers' onboard laboratories.

present in the soil. Samples were incubated for five days in an atmosphere of labeled gas, contained in a chamber illuminated by a xenon lamp to simulate sunlight. After incubation, the chamber was flushed of gas, so that any remaining radioactivity would have to be associated with the soil. The soil sample was then heated to 625°C (1157°F) to drive off volatile material (including any possible organic compounds), and the quantity of ^{14}C released was measured by detectors. Some accumulation of the radioactive label was detected, but this result was also found in the sterilized control: again, there was nothing to suggest biological processes as opposed to soil chemistry.

The labeled release experiment gave some confusing results. There was an initial rapid production of radioactive gas, which then remained constant over the 10-day period of the experiment. This is unlike the pattern of gas release that might be expected from microbial growth, where it would be normal to observe a lag interval after the addition of the nutrient medium, followed by a steady rise in gas production as the culture began to grow logarithmically, falling off again as nutrients became depleted. The sterilized control sample in this experiment did not show the pulse of labeled gas release. The most logical explanation for these results is that some non-biological component in the Martian soil was responsible, and that this component was destroyed in the sterilization heating.

Overall, there was a considerable weight of evidence against the existence of living organisms in the soil at the Viking landing sites. Critically, the gas chromatograph mass spectrometer (GCMS) instrument, capable of detecting traces down to one part in a billion, found no indication of organic material. At this level of sensitivity, whatever was reacting in the gas exchange experiment could not have been a living organism.

Organic material should have been present in the Martian soil, having been delivered by the continuing "rain" of meteorites over the ages. However, none could be detected by the Viking instruments. In the absence of an ozone layer, ultraviolet radiation from the Sun destroys organic molecules on the Martian surface fairly quickly: no microorganisms could survive long on the hostile environment of the Martian surface.

The Viking landers also excavated samples from underneath rocks, where they would have been shielded from ultraviolet. Again, the results proved negative for the existence of organic compounds, let alone biological activity.

The possibility that the labeled release experiment did show evidence for life has its supporters, notably Dr. Gil Levin, the experiment's designer. While most workers feel that the Viking results proved negative, some doubters claim that the GCMS detectors were insufficiently sensitive to rule out the presence of bacteria or organic material in the soil.

It is perhaps a fair argument that the Viking landers sampled the Martian soil at only two locations, which may not have been completely representative. In time, investigation of other sites will surely clarify the situation, starting in late 2003 with the Beagle 2 lander. Some scientists have suggested that searches in Mars' polar ice might be more fruitful.

Given the apparent absence of even traces of organic matter, however, the current consensus has to be that there is no compelling evidence for the existence of even simple life on Mars at this epoch.

Mars on Earth

Those who hope that we will, eventually, find microbial life on Mars take heart from the surprising resilience of terrestrial bacteria, some of which fill hostile ecological niches such as ocean floor hydrothermal vents or alkaline salt flats. In Antarctica, pink snowfields are sometimes found, their color resulting from localized growth of red-pigmented photosynthetic algae.

Antarctica's dry valleys, some of which have seen no precipitation in centuries, provide researchers with an environment quite similar in some respects to prevailing conditions on Mars. Photosynthetic bacteria have been found here, surviving the cold, arid conditions inside translucent rocks, and deriving their energy from the wan sunlight that filters through to them for part of the year. Perhaps there are simple organisms on Mars that survive the lethal solar ultraviolet radiation by the same expedient, and by searching in the surface soil we have been looking in the wrong place? The Mars Exploration Rovers might offer some clues in early 2004.

Antarctic meteorites

In addition to its similarities of aridity and low temperature, the Antarctic has played a key role in another part of the Martian life debate. The frozen continent has proved an incredibly rich source of meteorites, including several that have been identified as Martian in origin. These meteorites have provided scientists with samples of the planet's surface for direct investigation.

Until fairly recently, most meteorites available for study had been found fairly soon after falling, often as a result of their atmospheric entry having been seen as a brilliant fireball meteor. Searching for meteorites in desert areas, relatively free of native rocks, has been productive in, for example, Australia's Nullabor Plain. The richest

▶ *Several meteorites of Martian origin have been found on Earth. Known as SNC meteorites, they contain trapped gases in proportions that match those found by the landers. This meteorite, EETA 79001, was found at Elephant Moraine in Antarctica.*

source of all, having yielded up tens of thousands of meteorites, has been the Antarctic ice.

Meteorites falling on the continent's interior are – over the course of thousands of years – carried by the flow of glaciers toward the fringes. In certain areas, however, the ice encounters underlying mountains, resulting in the formation of "eddies" in which meteorites accumulate. Distinguished as areas of "blue ice," these are regions where ice is ablated by winds whipping across the glacier surface. As ice is ablated away, the meteorites trapped within it are exposed. Among the regions where meteorites can be picked up from the Antarctic ice in large numbers are Elephant Moraine, Yamoto Mountains and Allen Hills.

Most of the meteorites that have been found on Earth originate from the asteroid belt between the orbits of Mars and Jupiter, being fragments from collisions. Thrown into Earth-crossing trajectories, these fragments were eventually swept up by Earth's gravity. At least 21 meteorites of lunar origin – comparable in composition to the samples returned by the Apollo astronauts – are known. There are also 18 meteorites of Martian origin, which are sometimes referred to as the SNC meteorites after the earliest examples (Shergotty, Nakhla and Chassigny).

Material from Mars can only reach Earth if a violent impact on to the Red Planet's surface ejects fragments at greater than the escape velocity of 5 km/s (3 miles/s). Experts in orbital dynamics have confirmed the possibility of such events. The isotopic composition of tiny pockets of gas trapped within the meteorites has been used to confirm their Martian origins, matching well with measurements from the Viking landers.

Most of the Martian meteorites collected are fairly young, and basaltic in nature; it is from these that we can infer volcanic activity on Mars as recently as 150 million years ago.

Undoubtedly the most celebrated Martian meteorite is a 1.93 kg object picked up in Antarctica's Allen Hills late in 1984. Studies of ALH 84001 opened up once more the debate as to whether life could have existed on Mars in the distant past, even if there is no good evidence for it at the present epoch.

ALH 84001

ALH 84001 was studied at the Johnson Space Center, Houston, which acted as the receiving station for the Apollo lunar samples and is the storage facility for many of the recovered Antarctic meteorites. Analysis of the cosmic ray "weathering" of the meteorite's surface showed that it had been in orbit around the Sun for about 16 million years, before being swept up by Earth 13,000 years ago.

ALH 84001 is an igneous (volcanic) rock, which solidified early in Martian history, 4.4 billion years ago. After it had solidified, an impact nearby on Mars' surface produced fractures, and the rock spent some time immersed in water. The water, containing carbon dioxide, seeped into the cracks in ALH 84001. Chemical reactions then led to the formation of carbonate deposits, and it is on these that much of the interest in the meteorite has focused. The deposits formed between 4 and 3.6 billion years ago.

The carbonate is in the form of tiny, yellow-orange globules, 50 μm (0.05 mm) in diameter. The globules are rimmed with layers of magnesium- and iron-rich minerals, including iron sulfide and magnetite. Detailed studies of the globules were made by David S. McKay and Everett K. Gibson, who pointed out that iron sulfide in

◄ *The most celebrated of all Martian meteorites is surely ALH 84001, found in Antarctica's Allen Hills in 1984. Some workers believe ALH 84001 to contain microfossils.*

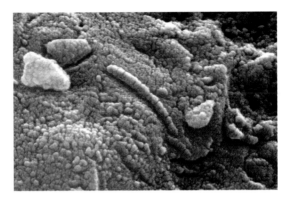

▶ *Worm-like structures in the ALH 84001 meteorite, revealed by electron microscopy, have been proposed by some researchers to be fossil Martian bacteria.*

terrestrial rocks is often associated with decaying organic matter. Organic material in the form of polycyclic aromatic hydrocarbons (PAHs), again commonly associated with decay, was also present in the ALH 84001 meteorite.

Tiny crystals of magnetite, 0.05 μm in size, were found in association with the carbonate globules. These, it is claimed by some scientists, are very similar to magnetite grains found in certain terrestrial bacteria. The crystals' linear arrangement might also be presented as possible evidence for biological origin.

Electron microscope images of material from the carbonate globules revealed "worm-like" structures, reminiscent of terrestrial rod-shaped bacteria. The size of the ALH 84001 rods is of the order of 20 to 100 nanometers (0.00002 to 0.0001 millimeter); for comparison, an *Escherichia coli* bacterium has a typical length of 2 μm, which is 200 times larger.

McKay and Gibson announced their findings in August 1996, and debate as to whether the structures they imaged really do represent fossil Martian bacteria has raged ever since, as have arguments surrounding the other suggested trace evidence for past life in ALH 84001.

Proponents of the idea point out that, at the time when the carbonate granules were deposited, Mars was a very different world from the one we see today – possibly even warm and wet, favoring biological activity. According to the fossil record, life had already appeared on Earth by this point in the Solar System's history and, as discussed above, nothing precludes a similar outcome on Mars. It has even been suggested by some that life on Earth could have been triggered by the arrival of a bacteria-bearing meteorite of Martian origin!

Most biologists are extremely sceptical. The alleged fossil bacteria are too small to have contained large biological molecules such as DNA; indeed, they are probably below the minimum sustainable size that a free-living biological structure must have if it is to function properly in enzyme-dependent activities like respiration. William Schopf, a leading authority on bacterial fossils, points out that even the smallest terrestrial examples – whose nature can be confirmed by comparison with types extant today in hostile environments on Earth – are far larger. Similarity of shape to bacilli is not sufficient to prove identity as fossil bacteria. It could as easily be claimed that the structures are an artifact resulting from preparation of the samples for electron microscopy in the first place.

Neither does the presence of PAHs have to be taken to imply that life once flourished inside ALH 84001. Polycyclic aromatic hydrocarbons are common – they are a by-product of barbecue cooking among other things – and have also been found in abundance among the organic molecule complement of carbonaceous chondrite meteorites. No-one would seriously propose PAHs in carbonaceous chondrites to be anything other than products of chemical – not biological – reactions.

The source of PAHs in ALH 84001 might also be debated. Although Antarctica has a relatively pristine environment, even here we can find traces of man-made industrial pollutants, including PAHs. As discussed above, the snows are sometimes colored by algal growth, which is another, natural, source of potential contamination. ALH 84001 could quite conceivably have acquired traces of organic material of terrestrial origin during its 13,000 years in or on the Antarctic ice.

Rather like the labeled release experiment aboard the Viking landers, the results from ALH 84001 are widely regarded as not providing evidence for life on Mars, but sufficient ambiguities remain for advocates of past Martian biology to continue to press their case.

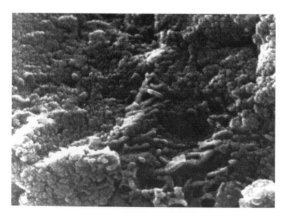

◀ Electron micrographs of the Martian meteorite ALH 84001 have revealed numerous structures similar in shape to terrestrial bacilli. Many biologists, however, consider the structures too small to be microfossils.

An absolutely definitive answer awaits further research, most particularly return of samples from Mars for which there can be no question of terrestrial contamination.

The question of whether life exists, or has ever existed, on Mars continues to rouse passions a century on from Percival Lowell and his alleged canals. Shining brightly in the skies of Earth in August 2003, November 2005 and December 2007, the Red Planet retains many mysteries even in the light of spacecraft exploration. Perhaps a generation from now, there will very definitely be life on Mars – as human explorers finally arrive there seeking answers to their never-ending questions about this intriguing neighbor world.

GLOSSARY

achromatic
A form of refractor in which combinations of optically different glasses are used in order to limit the effects of chromatic aberration.

albedo
Reflectivity. Dark features are described as having low albedo, while Mars' polar ice caps have high albedo. Mars' surface markings, being revealed by their subtle differences in reflectivity, are commonly referred to as albedo features.

altitude
The elevation, in degrees, of an astronomical body above the observer's horizon.

aperture
The clear diameter of a telescope's main, light-gathering objective lens or mirror.

aphelion
The point in its elliptical orbit at which a planet is farthest from the Sun.

apochromatic
A form of refractor, usually rather expensive, in which chromatic aberration should be almost totally absent thanks to the use of special glasses.

apparition
The visibility period of a planet.

appulse
The close approach, in line of sight, of two astronomical bodies.

arcsecond
An angular measure used to describe apparent distances on the sky. One arcsecond is a sixtieth of an arcminute, while there are 60 arcminutes in a degree.

arcminute
An angular measure used to describe apparent distances on the sky. One arcminute contains 60 arcseconds, while there are 60 arcminutes in a degree.

astronomical unit (AU)

The mean distance between Earth and the Sun. One astronomical unit is equal to 149,597,970 km (92,960,000 miles). It is a convenient way to describe distances in the Solar System.

celestial equator

A projection on to the celestial sphere of Earth's equator.

celestial sphere

As a matter of convenience for defining positions of astronomical bodies, it is convenient to think of these as appearing on the under-surface of an infinitely vast sphere surrounding the Earth. Positions on this celestial sphere are defined in terms of right ascension and declination.

central meridian

The line of longitude (connecting the north and south poles of a planet's disk) presented on the center of a planet as viewed from Earth.

chaotic terrain

Regions on Mars where ground collapse has produced canyons, landslides and other geological features characteristic of sudden local fracturing in the crust.

chromatic aberration

An optical effect found particularly in cheap refractors and eyepieces, where different wavelengths of light come to focus at different positions, resulting in undesirable color fringes around objects under view.

collimation

Optical alignment.

conjunction

The time at which two celestial bodies share the same right ascension in the sky. Planets are commonly described as being at conjunction when they are lost in the Sun's glare. It is also possible for planets to appear at conjunction with one another, relatively close together in line of sight in the sky.

core

In a terrestrial planet, the dense, central region. It is believed to consist of nickel-iron, which separated early in the planet's history while most of the body was still molten. Being heavier than other material, these dense metals sank to the center of the planet under gravity.

crust
The solid outer layer of a planet.

culmination
The point in a star's or planet's apparent path across the sky (caused by Earth's rotation) at which it lies on the meridian. It is usually taken to be the point at which the object is at its highest above the horizon.

declination
Celestial coordinate equivalent to latitude, expressed in degrees and minutes of arc. Objects have a positive declination if north of the celestial equator, and negative if to its south.

degree
Angular measure used to describe apparent distances on the sky. A great circle spanning the entire circumference of the celestial sphere contains 360 degrees of arc. The Moon subtends a diameter of about half a degree.

ecliptic
A great circle on the celestial sphere which marks out the Sun's apparent annual path relative to the "fixed" star background. Since their orbits lie close to its plane, the planets are usually found close to the ecliptic.

ejecta
Material excavated during impact, thrown out radially to form a "blanket" surrounding the resulting crater.

ephemeris (plural ephemerides)
A table listing positional, magnitude and other data for a celestial body as a function of date. Ephemeris data are normally presented for 00h UT (Universal Time) on the date shown.

focal length
The distance between the objective lens or mirror in a telescope and the point at which light is brought to a focus.

focal point
The position at which a lens or mirror brings light to a sharp focus.

focal ratio
The focal length of a telescope divided by its aperture.

gibbous
Term used to describe the phase of the Moon, or Mars, between half and full. The gibbous Moon is sometimes described, appositely, as almond-shaped.

great circle
A circle on the celestial sphere which has the full circumference of the sphere. The celestial equator and ecliptic are great circles.

inferior planet
A planet whose orbit lies closer to the Sun than that of Earth. Mercury and Venus are inferior planets.

limb
The visible edge of an extended astronomical body such as the Moon or Mars.

magma
Molten rock.

magnitude
In astronomical terminology, brightness. The smaller the numerical value, the brighter the object. When particularly bright, planets like Mars can have negative magnitude values.

mantle
In a terrestrial planet, the semifluid region between the core and crust. Reservoirs of magma in the mantle can break through the crust at weak points to produce volcanic activity; this certainly happens on Earth at present, and has occurred – possibly until quite recently – on Mars.

objective lens
In a refracting telescope, the main, "front-end" light-collecting lens.

occultation
An event in which one celestial body appears to pass in front of another. The Moon frequently occults stars. Such events are rarer for Mars, but have been observed. Mars itself can occasionally be occulted by the Moon.

opposition
The point at which a planet appears 180 degrees from the Sun in Earth's sky. At this time, the planet will be at its closest for the apparition, and will be visible throughout the hours of darkness.

perihelion
The point in its elliptical orbit at which a planet is closest to the Sun.

planitia
In standard planetary nomenclature, a plain.

protosolar nebula
Localized cloud of gas and dust which underwent gravitational collapse to form the Sun and planets 4.6 billion years ago.

planetesimal
One of countless small bodies that came together as a result of collisions early in Solar System history, resulting in formation of planets. Asteroids and comets are remnant rocky and icy planetesimals respectively.

protoplanet
A planet in the process of formation, by gravitational accumulation of planetesimals.

phase
For the Moon or a planet, its apparent degree of illumination. Marked changes in phase are seen for the Moon and inferior planets, ranging from Full, to gibbous, to half-phase and crescent then New. Of the superior planets, only Mars shows a marked phase, appearing gibbous when close to quadrature.

quadrature
The point in the orbit of a superior planet when it appears 90 degrees east or west of the Sun as seen from Earth.

refractor
A telescope that uses lenses to collect and focus light.

reflector
A telescope in which mirrors are used to collect and focus light.

resolving power
The ability of a telescope to separate closely spaced details. Resolving

power is governed in the first instance by aperture: larger apertures can resolve finer detail on Mars. However, resolving power is often limited by atmospheric seeing conditions.

right ascension
Celestial coordinate equivalent to longitude, increasing eastward from the position of the vernal equinox where the ecliptic and celestial equator intersect. Commonly abbreviated to RA, right ascension is described in hours, arcminutes and arcseconds. One hour of RA spans 15 degrees of sky.

rima
In standard planetary nomenclature, a fissure.

seeing
The steadiness of the air through which an astronomical object is viewed. Under conditions of good seeing, there are few air tremors and the image does not shimmer. Seeing should not be confused with transparency, the clarity of the air.

superior planet
A planet whose orbit lies outside that of Earth from the Sun; Mars, Jupiter, Saturn, Uranus and Neptune are all superior planets.

tectonics
Broad description of the processes occurring in the mobile crust of Earth. Earth's crust is made up of several interlocked segments – plates – which slide past, over or under each other over geological time. Mars' crust appears to be a single, rigid, immobile unit.

terminator
The day–night line on a planet or the Moon.

terrestrial planet
A planet, like Earth or Mars, which is composed principally of rocky materials.

Universal Time (UT)
Equivalent to Greenwich Mean Time. The standard time-system used in astronomical recording and reporting.

valles
In standard planetary nomenclature, a valley or canyon.

APPENDIX

BOOKS AND MAGAZINES

Astronomy Magazine. Box 1612, Waukesha, WI 53187 (http://www.astronomy.com).

Beatty, J.K., Petersen, C.C., and Chaikin, A. *The New Solar System* (4th ed.). Cambridge, UK: Cambridge University Press, 1999.

Dickinson, T. *NightWatch, A Practical Guide to Viewing the Universe*. Toronto: Firefly Books, 2001.

Dickinson, T., and Dyer, A. *The Backyard Astronomer's Guide*. Toronto: Firefly Books, 2002.

Harrington, P. *Star Ware*. New York: Wiley, 2002.

Norton, O.R. *The Cambridge Encyclopedia of Meteorites*. Cambridge, UK: Cambridge University Press, 2002.

Price, F.W. *The Planet Observer's Handbook*. Cambridge, UK: Cambridge University Press, 1994.

Scagell, R. *Guide to Stargazing with Your Telescope*. Cambridge, UK: Cambridge University Press, 2000.

Sheehan, W.H. *The Planet Mars: A History of Observation and Discovery*. Tucson: University of Arizona Press, 1996.

Sheehan, W.H., and O'Meara, S.J. *Mars: The Lure of the Red Planet*. New York: Prometheus Books, 2001.

Sky & Telescope Magazine. Box 9111, Belmont, MA 02178 (http://www.skypub.com).

SOFTWARE

TheSky (Software Bisque: http://www.bisque.com).

RedShift (Maris Multimedia Inc.: http://www.redshift.maris.com).

3D Atlas of Mars (Xamba Press: http://www.xamba.com).

CONTACTS

ALPO Mars homepage:
http://www.lpl.arizona.edu/~rhill/alpo/mars.html

Association of Lunar and Planetary Observers (ALPO). Executive Director: Donald C. Parker, 12911 Lerida Street, Coral Gables, FL 33156

Atlas of Mars: http://ic-www.arc.nasa.gov/ic/projects/bayes-group/Atlas/Mars

BAA Mars Section: http://marswatch.tn.cornell.edu/baa.html

Beagle 2 homepage: http://www.beagle2.com/index.htm

British Astronomical Association (BAA). Burlington House, Piccadilly, London, W1V 9AG, UK. (http://www.britastro.org/main/index.html)

Mars 2001 Odyssey homepage (THEMIS images):
http://themis.la.asu.edu

Mars Exploration Rover Mission homepage:
http://mars.jpl.nasa.gov/mer

Mars Express homepage: http://sci.esa.int/marsexpress

Mars Fact Sheet:
http://nssdc.gsfc.nasa.gov/planetary/factsheet/marsfact.html

Mars Global Explorer homepage: http://mars.jpl.nasa.gov/mgs

Mars Meteorites homepage: http://www.jpl.nasa.gov/snc

The Mars Society: http://marssociety.org

MarsNet: http://astrosun.tn.cornell.edu/marsnet/mnhome.html

The Planetary Society: http://www.planetary.org

QuickCam and Unconventional Imaging Astronomy Group (QCUIG): http://www.qcuiag.co.uk

INDEX

Page numbers in *italic* refer to captions.

ACKNOWLEDGEMENTS

Title NASA/JPL/USGS
Contents *center*
NASA/JPL/Malin Space Science
Systems; *sides*
NASA/JPL/Arizona State
University
Foreword NASA/JPL

Chapter 1
9 David Crisp and the WFPC
2 Science Team. JPL/CIT/NASA
10 NASA/JPL/Malin Space
Science Systems
11 NASA/JPL
14 NASA/JPL
15 NASA/JPL/Malin Space
Science Systems
16–17 Neil Bone

Chapter 2
19 Philip's
21 Philip's
23 Philip's
24 GSFC/NASA
26 *upper* NASA/JPL/Malin
Space Science Systems
26 *lower* NASA/JPL
27 NASA/JPL
28 NASA/JPL
30 NASA/JPL
31 *upper* NASA/JPL
31 *lower* USGS/NASA/JPL
32 NASA/JPL
33 NASA/JPL/Malin Space
Science Systems
34 *upper* NASA/JPL
34 *lower* Neil Bone
36 NASA/JPL

Chapter 3
40 Philip's
42 Neil Bone
45 Orion Optics
46 Meade Instruments Corp.
47 Philip's
59 Neil Bone
61 Damian Peach
62 Dr S.J. Wainwright

Chapter 4
64 R.J. McKim
66 Philip's
67 R.J. McKim
68 Damian Peach
69 R.J. McKim
70–71 Philip's/Paul Doherty
72 R.J. McKim
73 R.J. McKim
74 R.J. McKim
75 Damian Peach
77 R.J. McKim
78 R.J. McKim
81 Damian Peach

Chapter 5
86 Philip's/Wil Tirion

Chapter 6
95 Philip's/Wil Tirion

Chapter 7
102 Philip's/Wil Tirion

Chapter 8
107 Philip's
109 R.J. McKim, BAA Mars
Section
112 R.J. McKim, BAA Mars

Section
113 R.J. McKim, BAA Mars
Section
114 R.J. McKim, BAA Mars
Section
116 Lowell Observatory
117 R.J. McKim, BAA Mars
Section
118 R.J. McKim, BAA Mars
Section
119 R.J. McKim, BAA Mars
Section

Chapter 9
122 NASA/JPL
123 NASA/JPL
124 NASA/JPL
125 Smithsonian Institution
126 NASA/JPL
127 NASA/JPL
128 NASA/JPL
129 NASA/JPL
130 NASA/JPL
131 NASA/JPL/Malin Space
Science Systems
132 NASA/JPL
133 NASA/JPL/Arizona State
University
134 NASA/JPL
135 ESA
136 All Rights Reserved Beagle
2. http://www.beagle2.com

Chapter 10
140 NASA/JPL
143 JPL
144 GSFC/NASA
145 GSFC/NASA
146 GSFC/NASA